Skills Worksheet

Directed Reading A

Section: Four States of Matter

MATTER: MOVING PARTICLES

Write the letter of the correct answer in the space provided.

_______ 1. What are solids, liquids, and gases examples of?
   a. plasmas
   b. states of matter
   c. atoms
   d. molecules

_______ 2. Which of the following is true of the particles that make up matter?
   a. They are in constant motion.
   b. They only move in solids.
   c. They only move in liquids.
   d. They only move in gases.

Match the correct description with the correct term. Write the letter in the space provided.

_______ 3. Particles vibrate in place.

_______ 4. Particles move independently.

_______ 5. Particles are loosely connected.

   a. solid
   b. liquid
   c. gas

SOLIDS

Write the letter of the correct answer in the space provided.

_______ 6. What stays the same in solids?
   a. shape and volume
   b. color and shape
   c. position and shape
   d. state and volume

LIQUIDS

_______ 7. Which statement is true of liquids?
   a. They have a fixed shape but not a fixed volume.
   b. They have a fixed shape and a fixed volume.
   c. They have a fixed volume but not a fixed shape.
   d. They do not have a fixed shape or volume.

| Directed Reading A *continued*

_______ **8.** What happens to the volume of a liquid poured into a larger container?
   a. It increases.
   b. It decreases.
   c. It stays the same.
   d. It increases and then decreases.

GASES

_______ **9.** Which of the following states of matter has no fixed shape or volume and does not conduct electricity?
   a. gas
   b. liquid
   c. solid
   d. plasma

_______ **10.** What is true of the particles of a gas?
   a. They are very close together.
   b. They are always far apart.
   c. They slide past each other.
   d. They move freely and collide randomly.

_______ **11.** Why can a small tank of helium gas fill many balloons?
   a. because the amount of empty space between the particles can change
   b. because helium is a liquid
   c. because balloons are solid
   d. because the particles of helium in the tank are farther apart than particles in the balloon

PLASMAS

_______ **12.** What is the most common state of matter in the universe?
   a. plasma
   b. solid
   c. liquid
   d. gas

_______ **13.** Which of the following properties of plasma is different from gases?
   a. no definite shape
   b. no definite volume
   c. particles move freely
   d. conducts electric current

_______ **14.** How can artificial plasmas be formed?
   a. by passing electric charges through solids
   b. by passing electric charges through liquids
   c. by passing electric charges through gases
   d. by passing electric charges through plasmas

Directed Reading A

Section: Changes of State
ENERGY AND CHANGES OF STATE

Write the letter of the correct answer in the space provided.

______ **1.** What kind of change is a change of state?
    a. a chemical change
    b. a change in the identity of a substance
    c. a physical and chemical change
    d. a physical change

______ **2.** Which of the following is NOT true?
    a. Particles move differently depending on state.
    b. Energy must be added or removed to change state.
    c. Particles have the same amount of energy in different states.
    d. Particles have different amounts of energy in different states.

Match the labels to the graphic. Write the letter in the space provided.

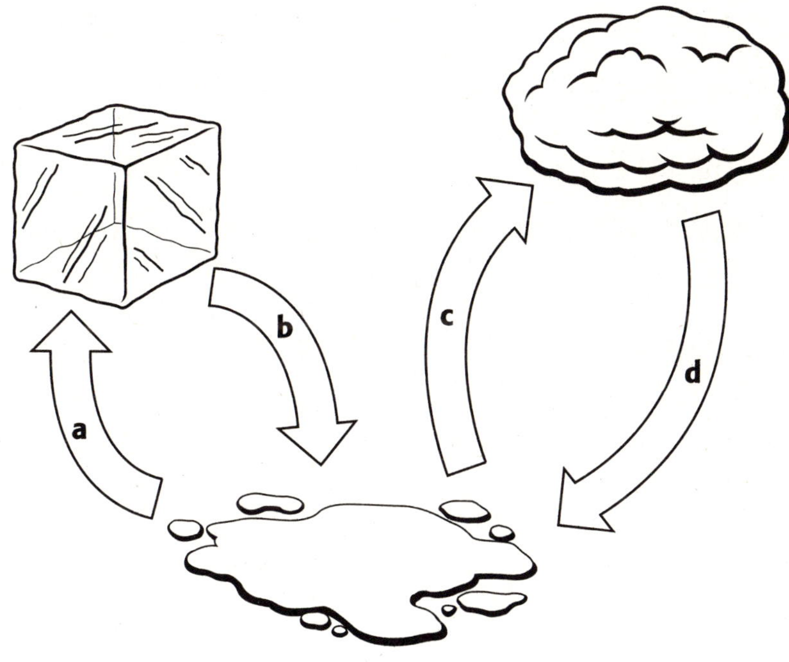

______ **3.** freezing

______ **4.** evaporation

______ **5.** condensation

______ **6.** melting

Directed Reading A *continued*

MELTING: SOLID TO LIQUID

Write the letter of the correct answer in the space provided.

______ **7.** What is a change in state from a solid to a liquid?
   a. freezing
   b. melting
   c. evaporation
   d. sublimation

Melting Point

______ **8.** What is the temperature at which a solid changes to a liquid called?
   a. melting point
   b. boiling point
   c. freezing point
   d. evaporation point

Adding Energy

______ **9.** What must be absorbed for a solid to melt?
   a. water
   b. atoms
   c. energy
   d. molecules

FREEZING: LIQUID TO SOLID

______ **10.** What is the change in state from a liquid to a solid called?
   a. evaporation
   b. melting
   c. sublimation
   d. freezing

Removing Energy

______ **11.** Which of the following is the same as the freezing point of an object?
   a. its melting point
   b. its evaporation point
   c. its sublimation point
   d. its boiling point

Directed Reading A *continued*

EVAPORATION: LIQUID TO GAS

_______ **12.** Which of the following is the change in state from a liquid to a gas?
  a. melting
  b. freezing
  c. sublimation
  d. evaporation

Evaporation and Boiling

_______ **13.** Which of the following is the conversion of liquid to vapor throughout the liquid?
  a. boiling
  b. freezing
  c. sublimation
  d. evaporation

_______ **14.** When water is boiling, which of the following does NOT happen?
  a. Molecular motion increases.
  b. Molecular motion decreases.
  c. Water molecules overcome attraction.
  d. Water vapor escapes.

Effects of Pressure on Boiling Point

_______ **15.** What happens to the boiling point of a substance as you go higher above sea level?
  a. The boiling point gets higher.
  b. The boiling point stays the same.
  c. The boiling point gets lower.
  d. The substance won't boil.

CONDENSATION: GAS TO LIQUID

_______ **16.** What is condensation?
  a. the change of state from a liquid to a gas
  b. the change of state from a solid to a gas
  c. the change of state from a gas to a solid
  d. the change of state from a gas to a liquid

_______ **17.** Which of the following must happen for gas to become a liquid?
  a. Small numbers of particles must clump together.
  b. Large numbers of particles must spread apart.
  c. Large numbers of particles must clump together.
  d. Small numbers of particles must spread apart.

| Directed Reading A *continued*

SUBLIMATION: SOLID TO GAS

______ **18.** Dry ice changes from a solid to a gas during what process?
- **a.** melting
- **b.** freezing
- **c.** boiling
- **d.** sublimation

______ **19.** What happens to the particles of a substance during sublimation?
- **a.** They spread far apart.
- **b.** They clump close together.
- **c.** They lose energy.
- **d.** They maintain the same energy.

TEMPERATURE AND CHANGES OF STATE

______ **20.** What happens when the temperature of a substance changes?
- **a.** The speed of the particles stays the same.
- **b.** The speed of the particles changes.
- **c.** The substance always melts.
- **d.** The substance always freezes.

Name _________________________________ Class _______________ Date __________

Directed Reading B

Section: Four States of Matter

MATTER: MOVING PARTICLES

1. What is a state of matter?

____

____

2. What are the three most familiar states of matter?

____

____

____

3. Matter is made up of particles called ___________________ and

___________________.

Match the correct description with the correct state of matter. Write the letter in the space provided.

_______ **4.** Particles do not move fast enough to overcome the strong attraction between them.

_______ **5.** Particles move independently of one another.

_______ **6.** Particles are close together but can slide past one another.

a. solid

b. liquid

c. gas

SOLIDS

_______ **7.** The particles of matter that make up a solid
  a. have a weaker attraction than those of a liquid.
  b. do not move at all.
  c. do not move fast enough to overcome the force of attraction.
  d. move from place to place.

8. What is the definition of a solid in terms of shape and volume?

____

____

____

LIQUIDS

9. How do the particles of a liquid make it possible to pour juice into a glass?

____

____

10. The juice in a beaker is poured into a graduated cylinder. The volume of juice in either container is 350 mL. What does this show you about the properties of a liquid?

____

____

GASES

11. What is the definition of a gas in terms of shape and volume?

____

____

12. How is it possible for one small tank of helium to fill hundreds of balloons?

____

____

____

PLASMAS

13. What state of matter makes up more than 99% of the matter in the universe?

____

14. How do plasmas behave differently than gases?

____

____

____

15. Give one example of a natural plasma and one example of an artificial plasma.

____

____

Name _________________________________ Class _______________ Date ___________

Directed Reading B

Section: Changes of State
ENERGY AND CHANGES OF STATE

_______ **1.** Which of the following have the most energy?
    a. particles in steam
    b. particles in liquid water
    c. particles in ice
    d. particles in freezing water

2. When a substance changes from one physical form to another, we say the

substance has undergone a(n) _____________________.

3. List the five main kinds of changes of state.

____

____

____

____

____

MELTING: SOLID TO LIQUID

4. Could you use gallium to make jewelry? Why or why not?

____

____

____

____

5. The temperature at which a substance changes from solid to liquid is

the _____________________ of the substance.

FREEZING: LIQUID TO SOLID

6. A substance's _____________________ is the temperature at which it

changes from a liquid to a solid.

| Directed Reading B *continued*

7. What happens if energy is added to or removed from a glass of ice water?

____

____

____

____

EVAPORATION: LIQUID TO GAS

Match the correct definition with the correct term. Write the letter in the space provided.

_______ **8.** the change of a substance from a liquid to a gas

_______ **9.** the change of state from a liquid to a gas when the vapor pressure equals the atmospheric pressure

_______ **10.** the temperature at which a liquid boils

a. boiling point

b. evaporation

c. boiling

11. As you go higher above sea level, the ___________________ decreases

and the ___________________ of a substance gets lower.

CONDENSATION: GAS TO LIQUID

12. The change of state from a gas to a liquid is called ___________________.

13. At a given pressure, the condensation point for a substance is the same as

its ___________________.

14. For a substance to change from a gas to a liquid, particles

must ___________________.

SUBLIMATION: SOLID TO GAS

15. Why is solid carbon dioxide called "dry ice"?

____

____

____

16. The change of state from a solid directly to a gas is called

____

Directed Reading B *continued*

TEMPERATURE AND CHANGES OF STATE

17. The speed of the particles in a substance changes when the

_______________________ changes.

18. When a substance is undergoing a change of state, the temperature of the

substance does not change until the _______________________ is complete.

Vocabulary and Section Summary A

Four States of Matter

VOCABULARY

In your own words, write a definition of the following terms in the space provided.

1. states of matter

____

____

2. solid

____

____

3. liquid

____

____

4. gas

____

____

5. plasma

____

____

SECTION SUMMARY

Read the following section summary.

- Particles of matter are in constant motion. The states of matter depend on the motion of particles.

- A solid has a definite shape and volume. A liquid has a definite volume but not a definite shape.

- A gas does not have a definite volume or shape. Plasma, a fourth state of matter, does not have a definite shape or volume, and its particles are broken apart.

Name _______________________________ Class _______________ Date ___________

Vocabulary and Section Summary A

Changes of State

VOCABULARY

In your own words, write a definition of the following terms in the space provided.

1. change of state

____

____

____

2. melting

____

____

____

3. evaporation

____

____

____

4. boiling

____

____

____

5. condensation

____

____

____

6. sublimation

____

____

____

Vocabulary and Section Summary A *continued*

SECTION SUMMARY

Read the following section summary.

- A change of state is the conversion of a substance from one physical form to another.
- A change of state requires a loss or gain of energy by a substance's particles.
- Melting is the change from a solid to a liquid, and freezing is the change from a liquid to a solid.
- Both boiling and evaporation result in a liquid changing to a gas.
- Condensation is the change of a gas to a liquid. It is the reverse of evaporation.
- Sublimation changes a solid directly to a gas.
- The temperature of a substance does not change during a change of state.

Vocabulary and Section Summary B

Four States of Matter

VOCABULARY

After you finish reading the section, try this puzzle! Unscramble the words to fill in the blanks below. Then, copy the numbered letters into the corresponding blanks at the bottom of the page to reveal what determines a substance's state of matter.

1. malpas

___ ___ ___ ___ ___ ___
 6   5   3

2. osslid

___ ___ ___ ___ ___ ___
       10   13

3. sagse

___ ___ ___ ___ ___
  7   2

4. siquldi

___ ___ ___ ___ ___ ___ ___
 11   4

5. tetsas fo taterm

___ ___ ___ ___ ___ ___ ___ ___ ___
  14 16 1 12

___ ___ ___ ___ ___ ___
   9   15 8

6. What determines a substance's state?

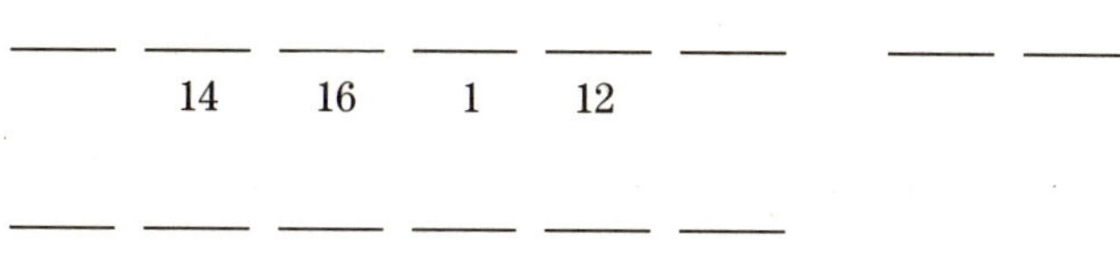

___ **H** ___ **W** ___ **Y** ___ ___ ___
 1   2   3   4 1 5

___ ___ ___ ___ ___ **C** ___ ___ ___
 6 7 8 9 10   11 12 13

___ **N** ___ ___ ___ ___ **C** ___
 10   14 15 8 16   14

Vocabulary and Section Summary B *continued*

SECTION SUMMARY

Read the following section summary.

- Particles of matter are in constant motion. The states of matter depend on the motion of particles.

- A solid has a definite shape and volume. A liquid has a definite volume but not a definite shape.

- A gas does not have a definite volume or shape. Plasma, a fourth state of matter, does not have a definite shape or volume, and its particles are broken apart.

Name _________________________________ Class _______________ Date _____________

Vocabulary and Section Summary B

Changes of State
VOCABULARY

After you finish reading the section, try this puzzle! Use the clues below to write the term described in the space provided. Then, find those words in the word search puzzle below. Terms can be hidden in the puzzle vertically, horizontally, diagonally, or backward.

_________________________ **1.** ice changing to water

_________________________ **2.** water vapor changing into rain

_________________________ **3.** dry ice changing into smoke

_________________________ **4.** occurs at a lower temperature in Denver, Colorado than at sea level

_________________________ **5.** water turning into steam

_________________________ **6.** a nonchemical change from one state to another

G	D	I	T	H	P	I	J	B	H	Y	P	N	A	X
G	K	G	T	O	A	W	P	A	O	K	O	X	B	S
N	R	W	I	S	E	B	J	L	S	I	K	J	F	N
O	P	Y	U	U	N	T	O	P	T	L	L	N	O	W
I	Y	A	F	Z	D	U	Y	A	V	B	Z	I	Y	W
T	O	I	S	O	N	I	M	Y	L	S	T	J	N	L
A	P	A	Z	G	V	I	V	X	G	A	G	Y	D	G
R	K	U	Y	K	L	U	U	A	S	F	U	K	Q	I
O	I	N	S	B	F	T	V	N	Y	J	M	Y	S	K
P	H	W	U	H	D	N	E	L	H	D	E	I	D	S
A	Z	S	B	P	B	D	K	W	O	P	L	N	Z	I
V	H	C	H	A	N	G	E	O	F	S	T	A	T	E
E	J	I	Y	O	Z	U	I	B	J	Z	I	Y	B	W
E	P	X	C	P	S	T	S	O	H	K	N	W	O	X
A	O	U	W	I	J	U	I	A	U	O	G	T	Y	T

▌Vocabulary and Section Summary B *continued*

SECTION SUMMARY

Read the following section summary.

- A change of state is the conversion of a substance from one physical form to another.
- A change of state requires a loss or gain of energy by a substance's particles.
- Melting is the change from a solid to a liquid, and freezing is the change from a liquid to a solid.
- Both boiling and evaporation result in a liquid changing to a gas.
- Condensation is the change of a gas to a liquid. It is the reverse of evaporation.
- Sublimation changes a solid directly to a gas.
- The temperature of a substance does not change during a change of state.

Reinforcement

Make a State-ment

Complete this worksheet after you finish reading the section "Four States of Matter."

Each figure below shows a container that is meant to hold one state of matter. Identify the state of matter, and write the state at the top of the figure's corresponding description box. Then, write each of the descriptions listed below in the correct boxes. Some descriptions may go in more than one box.

Particles are close together.

Particles are held tightly in place by other particles.

Particles break away completely from one another.

changes volume to fill its container

changes shape when placed in a different container

Amount of empty space can change.

has definite shape

Particles vibrate in place.

does not change in volume

State of Matter	Description

Critical Thinking

What a State!

From the *Journal of Galactic Research:*

Amazing Discovery of New Planet

Nobel Prize-winning astrophysicist Dr. Philo Philosophus has announced the discovery of a new planet named Phazon. Dr. Philosophus reports that although Phazon resembles Earth from a distance, it is really quite different. Matter on Phazon exists in three states—liquid, solid, and gas—as it does on Earth. On Phazon, however, each of these states of matter has one unique property.

Unique Properties of Matter on Planet Phazon

Solid	At high temperatures, solids always sublime from solid to gas.
Liquid	Liquids have no fixed volumes and must be stored in pressurized containers.
Gas	Gases have fixed volumes, as liquids on Earth do.

APPLYING CONCEPTS

1. Imagine that you have been chosen to visit Phazon. Do you think you will need special equipment to be able to breathe on the planet's surface? Explain your reasoning.

   ____

   ____

   ____

   ____

   ____

2. Temperatures on Phazon can be quite low. If wood were available, would it be possible to make a fire for warmth? Explain.

   HELPFUL HINT
   A fire needs oxygen in order to burn.

   ____

   ____

   ____

   ____

Critical Thinking *continued*

DEMONSTRATING REASONED JUDGMENT

3. The human body is composed mainly of liquids. Considering this fact, do you think it would be safe to visit planet Phazon without protective clothing? Explain your answer.

____

____

____

____

____

4. Give two examples of how cars on Phazon would differ from cars on Earth. Explain your reasoning.

____

____

____

____

____

____

SciLinks Activity

SOLIDS, LIQUIDS, AND GASES

Go to www.scilinks.org. To find links related to solids, liquids, and gases, type the keyword HY71420. Then, use the links to answer the following questions about solids, liquids, and gases.

> **Internet Resources**
>
> For a variety of links related to this chapter, go to www.scilinks.org
> Topic: Solids, Liquids, and Gases
> SciLinks code: HY71420

1. What are some characteristics of solids?

____

____

2. What are some characteristics of liquids?

____

____

3. What are some characteristics of gases?

____

____

In the diagram below, compare solids, liquids, and gases. Where two or more circles overlap, draw an arrow to the area of overlap, and list the ways that those states of matter are alike. Where the circles do not overlap, list the ways that each of these states of matter is unique.

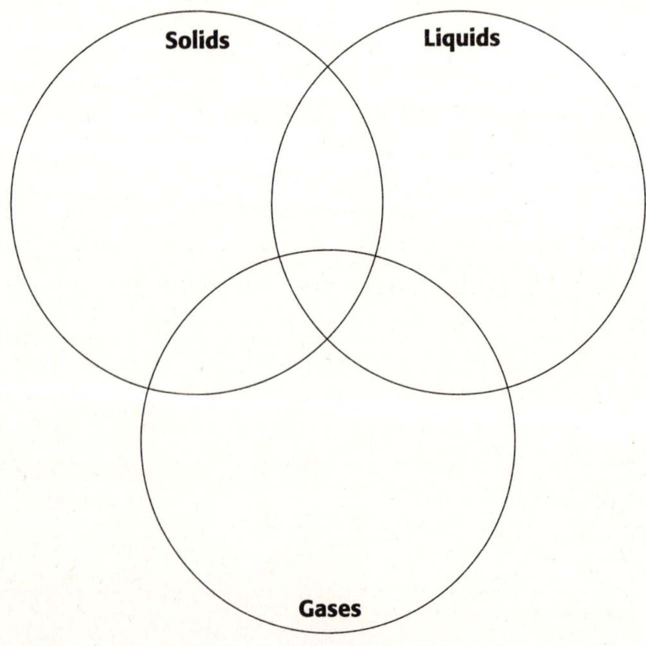

Section Review

Four States of Matter

USING VOCABULARY

1. Write an original definition for *gas* and *plasma*.

____

____

UNDERSTANDING CONCEPTS

2. Describing Describe the difference in particle motion between solids, liquids, and gases.

____

____

____

INTERPRETING GRAPHICS

Use the image below to answer the next two questions.

3. Identifying Identify the state of matter shown in the jar.

____

____

4. Concluding Can the individual particles inside the jar move? If so, describe how they move.

____

____

____

Section Review *continued*

CRITICAL THINKING

5. Analyzing Processes The volume of a gas can change, but the volume of a solid cannot. Explain in terms of particles why this is true.

____

____

____

____

____

____

____

CHALLENGE

6. Evaluating Hypotheses Tommy is planning an experiment to explore particle motion for two different substances. He hypothesizes that particle motion will be the same for each substance under the same conditions. What important factor has he not taken into account?

____

____

____

____

____

____

____

____

____

Skills Worksheet)

Section Review

Changes of State

USING VOCABULARY

For each pair of terms, explain how the meanings of the terms differ.

1. *boiling* and *melting*

____

____

2. *condensation* and *evaporation*

____

____

UNDERSTANDING CONCEPTS

3. Describing Describe how the motion and arrangement of particles in a substance change as the substance freezes.

____

____

4. Comparing How are boiling and evaporating similar? How are they different?

____

____

CRITICAL THINKING

5. Making Inferences Imagine that bubbles begin to form in a sample of liquid, but the temperature did not change. What must have happened to cause this change?

____

____

6. Analyzing Ideas When a solid melts, its density does not change very much. So, why do a liquid and a solid have such different physical properties?

____

____

Section Review *continued*

INTERPRETING GRAPHICS

Use the two pictures below to answer the next question.

 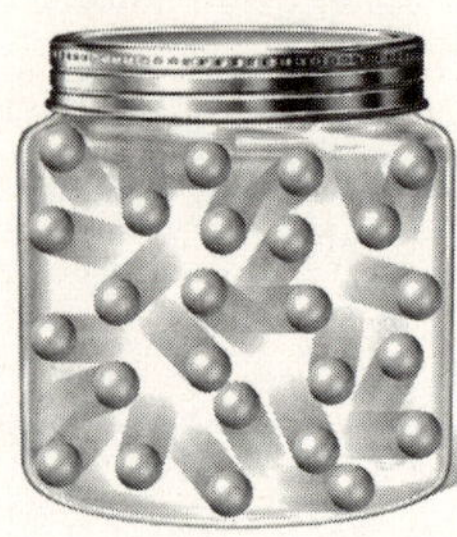

7. Analyzing Processes Describe two ways by which the particles in the picture above on the left could end up like the particles in the picture above on the right.

____

____

____

____

____

____

MATH SKILLS

8. Making Calculations If the volume of a substance in the gaseous state is 1,000 times the volume of that substance in the liquid state, how much space would 18 mL of that substance in the liquid state take up if it evaporated? Show your work below.

Name ____ Class _______________ Date ______________

Chapter Review

USING VOCABULARY

______ **1. Academic Vocabulary** Which of the following words means "a set of steps or events"?
   a. reaction
   b. process
   c. principle
   d. role

For each pair of terms, explain how the meanings of the terms differ.

 2. *solid* and *liquid*

____

____

 3. *evaporation* and *boiling*

____

____

 4. *condensation* and *sublimation*

____

____

UNDERSTANDING CONCEPTS
Multiple Choice

______ **5.** Which of the following statements best describes the particles of a liquid?
   a. The particles are far apart and moving fast.
   b. The particles are close together but moving past each other.
   c. The particles are far apart and moving slowly.
   d. The particles are closely packed and vibrating in place.

______ **6.** Dew collecting on a spider web in the early morning is an example of
   a. condensation.
   b. evaporation.
   c. sublimation.
   d. melting.

❚ Chapter Review *continued*

_______ **7.** During which change of state do atoms or molecules become more
  ordered?
  a. boiling
  b. condensation
  c. melting
  d. sublimation

_______ **8.** As the particles of a solid undergo sublimation, they
  a. lose energy.
  b. move closer to one another.
  c. change temperature.
  d. move farther apart from one another.

Short Answer

9. Listing Rank solids, liquids, and gases in order of particle speed from the
highest speed to the lowest speed.

____

____

10. Classifying At atmospheric pressure, what is the characteristic boiling point
of water, in degrees Celsius?

____

11. Analyzing Explain why liquid water takes the shape of its container but an ice
cube does not.

____

____

12. Concluding Water's states of matter include steam, liquid water, and ice. What
about water is the same in these states? What can you conclude about what
changes and what does not change during a change of state?

____

____

____

____

____

____

| Chapter Review *continued*

INTERPRETING GRAPHICS

Use the graph below to answer the next two questions.

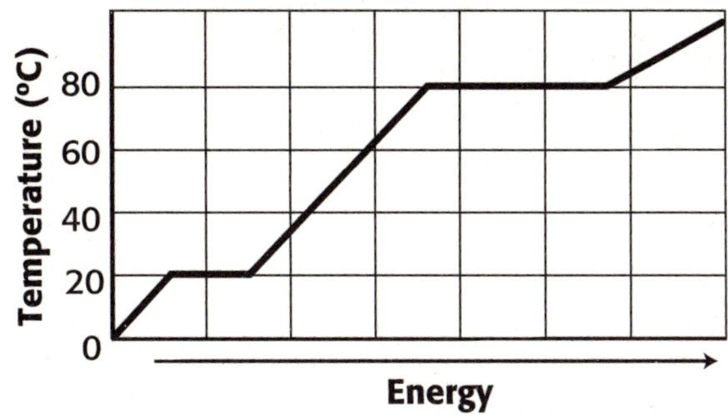

13. Identifying What is the boiling point of the substance? What is the melting point?

____

14. Concluding How does the substance change while energy is being added to the liquid at 20°C?

____

____

WRITING SKILLS

15. Creative Writing Imagine that you are a gas particle and that the material you are in condenses and then freezes. From your point of view as a particle, write a clear step-by-step description of what happens as you go through each change of state.

____

____

____

____

____

____

____

____

____

____

CRITICAL THINKING

16. Concept Mapping Use the following terms to create a concept map: *states of matter, solid, liquid, gas, changes of state, freezing, evaporation, condensation,* and *melting.*

Chapter Review *continued*

INTERPRETING GRAPHICS

Use the picture below to answer the next two questions.

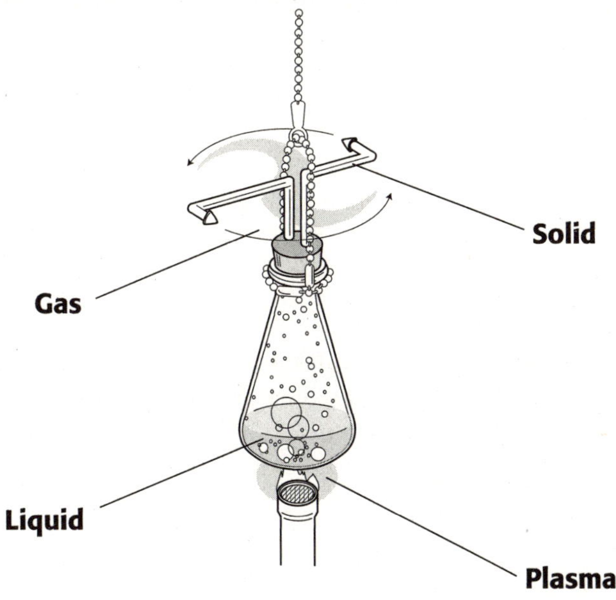

17. Analyzing Processes Explain, based on the motion of the gas particles, how the gas is causing the tubes to spin around as shown in the diagram above.

____

____

____

____

____

18. Making Inferences If the liquid shown in the picture above is water and if the air pressure is equal to the normal atmospheric pressure at sea level, what must the temperature of the water be, in degrees Celsius?

____

____

____

____

____

Chapter Review *continued*

19. Applying Concepts After taking a shower, you notice that small droplets of water cover the bathroom mirror. Explain how these drops form. Be sure to describe where the water comes from and the changes it undergoes.

____

____

____

____

20. Making Inferences At sea level, water boils at 100°C and methane boils at −161°C. Which of these substances has a stronger force of attraction between its particles? Explain your reasoning.

____

____

____

____

21. Analyzing Ideas By using an electric current, you can split liquid water to form two new substances, hydrogen and oxygen gases. Is this a change of state? Explain your answer.

____

____

____

22. Evaluating Hypotheses Imagine that a gas is bubbling up from a sample of water. Laurel forms the hypothesis that the water is boiling. How could she test that hypothesis?

____

____

____

____

MATH SKILLS

23. Analyzing Data Kate placed 100 mL of water in five different pans, placed
the pans on a windowsill for a week, and measured how much water
evaporated from each pan. Draw a graph of her data, which is shown below.
Place surface area on the x-axis and volume evaporated on the y-axis.
Is the graph linear or nonlinear? What does this information tell you?

Pan number	1	2	3	4	5
Surface area (cm²)	44	82	20	30	65
Volume evaporated (mL)	42	79	19	29	62

____

____

____

CHALLENGE

24. Analyzing Methods To protect their crops during freezing temperatures,
orange growers spray water onto the trees and allow it to freeze. In terms of
energy lost and energy gained, explain why this practice protects the oranges
from damage.

____

____

____

____

____

Chapter Pretest

Teacher Notes and Answer Key

The Pretest questions are designed to help you determine the prior knowledge of your students. Some questions test whether students have mastered the background knowledge they need to understand the content you are about to teach. Other questions test your students' prior knowledge of the content you are about to teach. Use the Pretest with the Test Doctors and diagnostic teaching tips in these notes pages to help you tailor your instruction to your students' specific needs.

QUESTION NUMBER	CORRECT ANSWER	STANDARD
1	A	8.5.a
2	B	8.7.c
3	A	8.8.a
4	C	8.8.b
5	B	8.8.b
6	B	8.3.d
7	D	8.3.e
8	A	8.5.d
9	B	8.7.c

TEST DOCTOR

The following Pretest questions have been diagnosed by the Test Doctor. Find out what might be causing your students' "ailing" answers. Each Test Doctor is followed by a diagnostic teaching tip to help you address students' learning needs.

Question 1 *asks students to identify a chemical change.*

A Correct. Rusting is an example of a chemical change.
B Incorrect. Crushing is an example of a physical change.
C Incorrect. Shaving is an example of a physical change.
D Incorrect. Liquefaction, or condensation, is an example of a physical change.

> **Diagnostic Teaching Tip:** Students who have difficulty distinguishing between a physical change and a chemical change might benefit from a review of these two concepts. Remind students that a physical change involves changing a physical property, such as shape or state. A chemical change occurs when atoms or molecules interact to form a new substance with new properties.

Question 2 *asks students to identify which property is NOT unique to an element.*

A Incorrect. Density is a unique property of an element.
B Correct. External temperature is not a unique property of an element.
C Incorrect. Malleability is a unique property of an element.
D Incorrect. Thermal conductivity is a unique property of an element.

Chapter Pretest *continued*

Diagnostic Teaching Tip: Students who answer this question incorrectly should review the properties of elements. Prepare a list of properties of elements for students. The list should include density, ductility, electrical conductivity, flexibility, magnetism, malleability, solubility, state, strength, and thermal conductivity. Invite volunteers to give a definition for each property.

Question 3 *asks students to identify the relationship between density and volume.*

A **Correct.** This is the definition of density.
B **Incorrect.** This is the inverse of density.
C **Incorrect.** For density, mass is divided by unit volume, not multiplied by it.
D **Incorrect.** Unit mass is not a consideration with density.

Diagnostic Teaching Tip: Students who have difficulty with this question might benefit from a review of writing mathematical equations from words. Write the correct equation for density. Then, go through each response with students, writing the "response" equation as you read. Point out the incorrect usage of the term *unit* with *mass* in the last response.

Question 4 *asks students to compute the density of silver, given its mass and volume.*

A **Incorrect.** This is the inverse of the density of silver.
B **Incorrect.** This is the given volume of the silver.
C **Correct.** This is the density of silver.
D **Incorrect.** This is the given mass of the silver.

Diagnostic Teaching Tip: Students who have trouble answering this question correctly should review the definition of density. Remind students that to calculate the density correctly, they must reduce the fraction so that it contains "1 cubic centimeter" as its denominator (unit volume).

Question 5 *asks students to compute the mass of a block of cork, given its volume and density.*

A **Incorrect.** This is the given density (with unit volume) of cork.
B **Correct.** This is the mass of the cork.
C **Incorrect.** This is a nonsensical value adding the length to the density.
D **Incorrect.** This is the given volume divided by the given density.

Diagnostic Teaching Tip: Students who have difficulty answering this question correctly might benefit from calculating the mass in a step-by-step process. First, write the equation for density. Rewrite the equation in terms of mass (mass equals density times volume). Ask a student volunteer to show how the volume of the block of wood was found (1 cm times 1 cm times 3 cm equals 3 cubic centimeters). Then, fill in the values for volume and density in the equation. Solve the equation for students.

| Chapter Pretest *continued*

Question 6 *asks students to determine which statement is true about states of matter.*

A Incorrect. The temperature of a substance may vary.

B Correct. Particles of a substance are always in motion.

C Incorrect. Particles of a liquid or gas may vary in their distance from each other.

D Incorrect. A liquid has no definite shape, and a gas has no definite shape or volume.

Diagnostic Teaching Tip: Students who have difficulty answering this question correctly might benefit from a demonstration showing the three states of matter. Show students an ice cube, a bottle partially filled with water, and a semi-blown-up balloon. Invite student volunteers to suggest differences and similarities between the three states of matter. Write "solid," "liquid," and "gas" on the board, and list the suggestions under the appropriate head. Be sure to emphasize Section 1, "Four States of Matter," in Chapter 4, "States of Matter." Students must understand states of matter (solid, liquid, gas) and molecular motion in order to master standard 8.3.d.

Question 7 *asks students to identify a state of matter based on the behavior of its particles.*

A Incorrect. In plasma, there is no definite shape or volume, and molecules have broken apart.

B Incorrect. In a gas, molecules move independently and collide with each other frequently.

C Incorrect. In a liquid, molecules are more loosely connected than those in a solid and can collide with or move past each other.

D Correct. In a solid, molecules vibrate and are packed together closely.

Diagnostic Teaching Tip: Students who have difficulty answering this question correctly might benefit from additional insight into the states of matter. Place the three examples used in the Diagnostic Teaching Tip for Question 6 in front of the class. Have students turn to Section 1, "Four States of Matter," in Chapter 4, "States of Matter," to the figure that shows models of a solid, a liquid, and a gas. Correlate each picture showing particle arrangement and movement to the concrete example. Be sure to emphasize Section 1, "Four States of Matter," in Chapter 4, "States of Matter." Students must know that in solids, the atoms are closely locked in position and can only vibrate, in order to master standard 8.3.e.

Question 8 *asks students to identify a proposition true of the term* freezing point.

A Correct. Freezing point is a physical property.

B Incorrect. Freezing point does not involve a chemical change (no new substance is formed).

C Incorrect. When freezing, a substance changes from a liquid to a solid.

D Incorrect. Only the freezing point of water is 0°C.

Diagnostic Teaching Tip: If students have difficulty with this question, prepare a glass bowl holding a mixture of crushed ice and water. Tell students that this is similar to a substance in the process of freezing. Go over each response with students, relating the response to the water/ice mixture. Tell students that although this particular mixture freezes at 0°C, not all substances freeze at that point. For example, antifreeze freezes at a lower temperature. Be sure to emphasize Section 2, "Changes of State," in Chapter 4, "States of Matter." Students must know that physical processes include freezing, in which a material changes form with no chemical reaction, in order to master standard 8.5.d.

Question 9 *asks students to determine the identity of a metal from a table, given certain properties.*

A **Incorrect.** Aluminum does not have a density of about 8.93 g/cm^3.
B **Correct.** Copper meets all of the given values.
C **Incorrect** Nickel has a hardness greater than 3.5.
D **Incorrect.** Silver does not have a density of about 8.93 g/cm^3.

Diagnostic Teaching Tip: For students who have difficulty answering this question correctly, have them team with another student. Have student pairs go over each given value. As each value is discussed, have them cross out any element that does not fit the value. Tell students that since melting point was not able to be determined, they may not use this value to help determine the element. Be sure to emphasize Section 2, "Changes of State," in Chapter 4, "States of Matter." Students must know that substances can be classified by their properties, including their melting temperature, density, and hardness, in order to master standard 8.7.c.

Chapter Pretest

______ **1.** Which of the following is an example of a chemical change?
   A rusting metal
   B crushed ice
   C shaving wood
   D liquefying nitrogen

______ **2.** All of the following are unique properties of an element EXCEPT for
   A density.
   B external temperature.
   C malleability.
   D thermal conductivity.

______ **3.** Which of the following is the correct relationship between density and volume?
   A Density is the ratio of the mass of a substance to the volume of the substance.
   B Density is equal to unit volume divided by mass.
   C Density is equal to mass times unit volume.
   D Density is equal to unit mass times volume.

______ **4.** A chunk of silver has a mass of 21.0 grams and a volume of 2 cubic centimeters. What is its density?
   A $1/10.5$ g/cm^3
   B 2 g/cm^3
   C 10.5 g/cm^3
   D 21.0 g/cm^3

______ **5.** A block of cork measures 1 cm by 1 cm by 3 cm and has a density of 0.25 g/cm^3. What is its mass?
   A 0.25 g
   B 0.75 g
   C 3.25 g
   D 12.0 g

______ **6.** Which of the following is true about all states of matter?
   A The temperature of a substance is always the same.
   B Particles of a substance are always in motion.
   C Particles of a substance are always the same distance apart from each other.
   D Matter always has a definite shape and volume.

Chapter Pretest *continued*

_______ **7.** A substance that contains vibrating molecules packed together closely
is probably
A a plasma.
B a gas.
C a liquid.
D a solid.

_______ **8.** Which of the following is true of a substance's freezing point?
A It is a physical property.
B It is a chemical change.
C The substance changes from a gas to a liquid.
D It only occurs at 0°C.

**The table below shows some properties of four different elements. Use the table
to answer the question that follows.**

ELEMENT	MELTING POINT	DENSITY	HARDNESS (MOH'S SCALE)
Aluminum	660.25°C	2.70 g/cm^3	2.75
Copper	1084.6°C	8.96 g/cm^3	3
Nickel	1453°C	8.9 g/cm^3	4
Silver	961°C	10.50 g/cm^3	2.5

_______ **9.** Josh and Melinda are trying to identify a mystery metal. They cannot
melt the metal in their laboratory. They calculate the metal's density to
be about 8.93 g/cm^3. They know its hardness is between 2.0 and 3.5.
They also know it is one of the metals shown in the table. Which metal
is it?
A aluminum
B copper
C nickel
D silver

Section Quiz

Section: Four States of Matter

Write the letter of the correct answer in the space provided.

_______ **1.** Which of the following statements is NOT true of atoms and molecules?
   a. They are tiny particles.
   b. They are always in motion.
   c. They are found in all matter.
   d. They never bump into each other.

_______ **2.** In a solid, the particles
   a. overcome the strong attraction between them.
   b. vibrate in place.
   c. slide past one another.
   d. move independently of one another.

_______ **3.** A gas
   a. has a definite volume but no definite shape.
   b. has a definite shape but no definite volume.
   c. has fast-moving particles.
   d. has particles that are always close together.

Match the correct description with the correct term. Write the letter in the space provided.

_______ **4.** consists of free-moving ions and electrons

_______ **5.** does not have a definite volume or shape, and does not conduct electricity

_______ **6.** includes the physical forms of matter

_______ **7.** has a fixed volume and shape

_______ **8.** has a definite volume but not a definite shape

a. states of matter

b. solid

c. liquid

d. gas

e. plasma

Section Quiz

Section: Changes of State

Match the correct definition with the correct term. Write the letter in the space provided.

_______ **1.** the change of state from a solid to a liquid

_______ **2.** the change of a substance from one physical form to another

_______ **3.** the change of state from a solid directly to a gas

_______ **4.** the change of state from a liquid to a gas

_______ **5.** the change of a liquid to a vapor throughout the liquid

_______ **6.** the change of state from a gas to a liquid

_______ **7.** the temperature at which a substance melts

_______ **8.** the change of state from a liquid to a solid

_______ **9.** the temperature at which a substance boils

a. change of state

b. melting

c. evaporation

d. boiling

e. condensation

f. sublimation

g. freezing

h. melting point

i. boiling point

Assessment

Chapter Test A

States of Matter
MULTIPLE CHOICE
Write the letter of the correct answer in the space provided.

_______ 1. Which of the following describes the tiny particles that make up matter?
   a. They are sometimes in motion.
   b. They are in constant motion.
   c. They are never in motion.
   d. They are sometimes not in motion.

_______ 2. Which of the following statements is NOT true?
   a. Particles of a gas move independently and collide frequently.
   b. Particles of a solid vibrate in place.
   c. Particles of a liquid can move past one another.
   d. Particles of a gas are locked in position.

_______ 3. What happens when a liquid becomes a gas?
   a. The particles give off energy.
   b. The particles move more quickly.
   c. The particles move closer together.
   d. The particles slow down.

_______ 4. Which of the following is true about a change of state?
   a. It is a chemical change.
   b. It is a chemical and physical change.
   c. It is a physical change.
   d. It can be either a physical or chemical change.

_______ 5. What is the same as the melting point of mercury?
   a. its boiling point
   b. its condensation point
   c. its freezing point
   d. its sublimation point

_______ **6.** Carbon dioxide gas enters the air from the surface of a block of dry
ice. What happens to the carbon dioxide in this process?
  a. The carbon dioxide gains energy.
  b. The carbon dioxide boils.
  c. The carbon dioxide increases in pressure.
  d. The carbon dioxide releases energy.

_______ **7.** Which of the following is true about a substance's melting point?
  a. It is the same as its boiling point.
  b. It is the same as its evaporation point.
  c. It varies depending on temperature.
  d. It is the same as its freezing point.

_______ **8.** Which of the following is a direct change from solid to gas?
  a. sublimation
  b. evaporation
  c. condensation
  d. melting

_______ **9.** Which of the following is NOT true about boiling?
  a. Particles throughout the liquid move faster.
  b. Only particles at the surface of a liquid can turn to gas.
  c. A substance's boiling point is the same as its condensation point.
  d. Particles throughout the liquid evaporate.

MATCHING

Match the correct description with the correct term. Write the letter in the space provided.

_______ **10.** This occurs whenever matter changes form.   **a.** states of matter

_______ **11.** Ice, water, and steam are examples of these.   **b.** change of state

     c. boiling

_______ **12.** This occurs when the vapor pressure of a
liquid equals the atmospheric pressure.

Chapter Test A *continued*

Match the correct description with the correct term. Write the letter in the space provided.

______ **13.** This requires the removal of energy from a gas.

______ **14.** This requires the addition of energy to a liquid.

______ **15.** This requires the addition of energy to a solid.

______ **16.** This requires the removal of energy from a liquid.

a. freezing

b. melting

c. evaporation

d. condensation

FILL-IN-THE-BLANK

Use the terms from the following list to complete the sentences below.

solid   liquid

plasma   gas

17. A state of matter with a fixed shape and volume is a(n)

________________________.

18. A state of matter that consists of free-moving ions and electrons is

________________________.

19. A state of matter with a fixed volume, but not a fixed shape,

is a(n) ________________________.

20. The particles of a(n) ________________________ are far apart.

Chapter Test B

States of Matter
MULTIPLE CHOICE

Write the letter of the correct answer in the space provided.

_______ **1.** Which of these factors could affect the temperature at which water boils?
   a. the volume of water in the pot
   b. the atmospheric pressure at which the water is heated
   c. the amount of energy added to the water
   d. the type of fuel used to heat the water

_______ **2.** How do the particles of water that evaporate from an open container differ from the particles that remain?
   a. The evaporated particles have only more speed.
   b. The evaporated particles have greater order.
   c. The evaporated particles have only higher energy.
   d. The evaporated particles have more speed and higher energy.

_______ **3.** Which of the following occurs when a liquid becomes a gas?
   a. The particles give off energy.
   b. The particles break away from one another.
   c. The particles move closer together.
   d. The particles slow down.

_______ **4.** The melting point of water is the same as its
   a. boiling point.
   b. condensation point.
   c. freezing point.
   d. sublimation point.

_______ **5.** A liter of gasoline will boil at
   a. a higher temperature than a milliliter of gasoline.
   b. a lower temperature than a milliliter of gasoline.
   c. the same temperature as a milliliter of gasoline.
   d. the same temperature as a milliliter of water.

_______ **6.** In order for carbon dioxide gas to enter the air from dry ice, particles in the dry ice must
   a. gain energy.
   b. boil.
   c. increase in pressure.
   d. undergo an exothermic change.

Chapter Test B *continued*

_______ **7.** Which of the following statements is NOT true of all different types of matter?
  a. They are made up of atoms and molecules.
  b. The particles that make them up are always in motion.
  c. They are made up of extremely small particles.
  d. The particles that make them up move at the same speed.

_______ **8.** A graph that shows the change in temperature of a substance as it is heated will show
  a. a straight line as the substance melts.
  b. a straight line as the substance freezes.
  c. a rising line as the substance melts.
  d. a falling line as the substance melts.

_______ **9.** The reverse of condensation is
  a. boiling.
  b. evaporation.
  c. freezing.
  d. sublimation.

_______ **10.** Solid dry ice changes to carbon dioxide gas through what change of state?
  a. melting
  b. evaporation
  c. freezing
  d. sublimation

MATCHING

Match the correct description with the correct term. Write the letter in the space provided.

_______ **11.** It starts as a gas and then becomes ionized.

_______ **12.** This happens when tomato soup boils.

_______ **13.** Solids, liquids, and gases are all examples.

_______ **14.** It has atoms and molecules that are close together but can slide past each other.

_______ **15.** It is the result of evaporation.

_______ **16.** It is the result of freezing.

a. states of matter
b. solid
c. liquid
d. gas
e. plasma
f. change of state

Use the figure below to answer questions 17 through 20. Write the letter of the correct answer in the space provided.

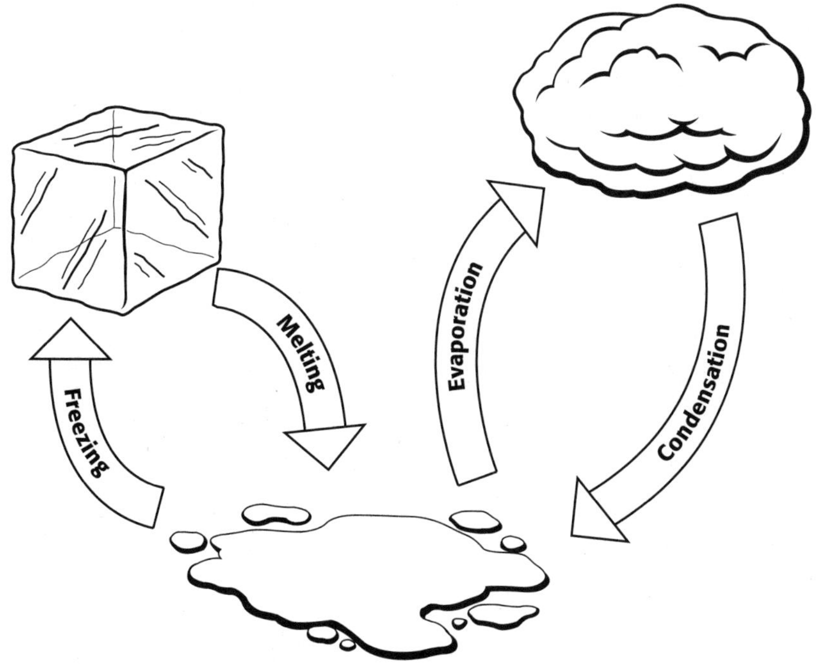

______ **17.** Which of the changes of state shown in the figure above require an increase in energy?
    a. freezing and evaporation   **c.** evaporation and melting
    b. freezing and condensation   **d.** condensation and melting

______ **18.** Which of the changes of state shown in the figure above require a decrease in energy?
    a. freezing and evaporation   **c.** evaporation and condensation
    b. condensation and freezing   **d.** melting and evaporation

______ **19.** Boiling water is an example of which of the changes of state shown in the figure above?
    a. freezing   **c.** evaporation
    b. condensation   **d.** melting

______ **20.** Which of the following is NOT true of the changes of state shown in the figure above?
    a. The melting point and freezing point are the same for a given substance.
    b. The temperatures at which condensation and evaporation occur are the same for a given substance.
    c. A solid cannot change to a gas through any of the changes of state shown above.
    d. A solid can change directly to a gas through evaporation.

Chapter Test C

States of Matter

USING KEY TERMS

Use the terms from the following list to complete the sentences below. Each term may be used only once. Some terms may not be used.

state of matter	condensation	evaporation
gas	liquid	sublimation
plasma	freezing	melting

1. The drops of water that appear on the outside of a glass of cold juice on a warm day are an example of _______________________.

2. The change of state from a liquid to a gas is _______________________.

3. Sublimation is a change of state from a solid directly to a(n) _______________________.

4. The motion of the particles of a substance determines its _______________________.

5. Melting is a change of state from a solid to a(n) _______________________.

6. A gas that has been ionized is called a(n) _______________________.

7. The change of state from a solid directly to a gas is called _______________________.

8. Heating a candle until the wax drips is an example of _______________________.

UNDERSTANDING KEY IDEAS

Write the letter of the correct answer in the space provided.

______ **9.** Which of the following examples involves the decrease of energy in a substance?
  a. ice melting in a glass of lemonade
  b. water boiling in a large pot
  c. gaseous water particles coming together to form droplets on a cup
  d. air in a bicycle tire gaining pressure on a hot day

______ **10.** Which of these factors could affect the temperature at which water boils?
  a. the size and shape of the pot in which the water is heated
  b. the atmospheric pressure at which the water is heated
  c. the amount of heat added to the water
  d. the temperature of the water before it is heated

| Chapter Test C *continued*

_______ **11.** Liquids tend to maintain a constant
      a. volume.   **c.** surface tension.
      b. pressure.   **d.** viscosity.

_______ **12.** Which of the following occurs when a gas becomes a liquid?
      a. The particles create energy.
      b. The particles break away from one another.
      c. The particles clump together.
      d. The particles stop moving.

13. What happens to the temperature of a pot of boiling water as the water evaporates?

____

____

____

14. A container with a mixture of water and ice is at 0°C. What will happen if energy is added to or removed from the water?

____

____

____

CRITICAL THINKING

15. Explain why more time is required to boil pasta in Denver, Colorado than in New Orleans, Louisiana.

____

____

____

____

16. To create special effects for movies and television shows, technicians often pour water over solid carbon dioxide, also called dry ice. What effect does this produce? Explain your answer.

____

____

____

____

Chapter Test C *continued*

17. On Thanksgiving Day, a big parade is held in New York City with many giant helium-filled balloons. How will the weather affect the inflating of the balloons? Explain your answer.

____

____

____

____

____

____

INTERPRETING GRAPHICS

The following graph shows the different states of water as it is cooled. Examine the graph, and answer the questions that follow.

Cooling Curve of Water

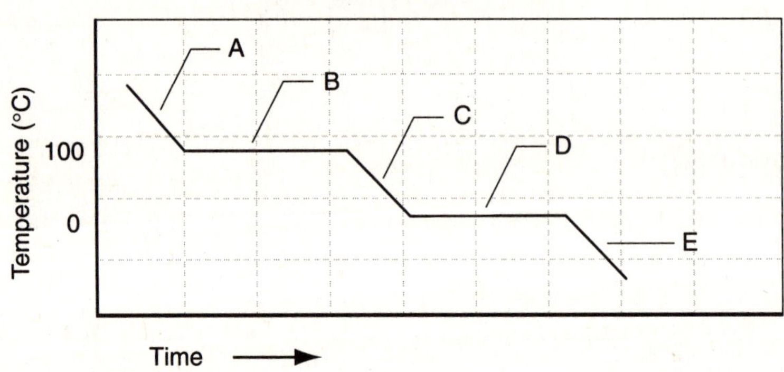

18. Which letters on the graph represent the three states of matter: gas, solid, and liquid?

____

____

19. Which letters on the graph represent the changes of state of condensation and freezing?

____

____

Chapter Test C *continued*

CONCEPT MAPPING

20. Use the following terms to complete the concept map below:

gases   condensation   states of matter
plasma   sublimation   change of state

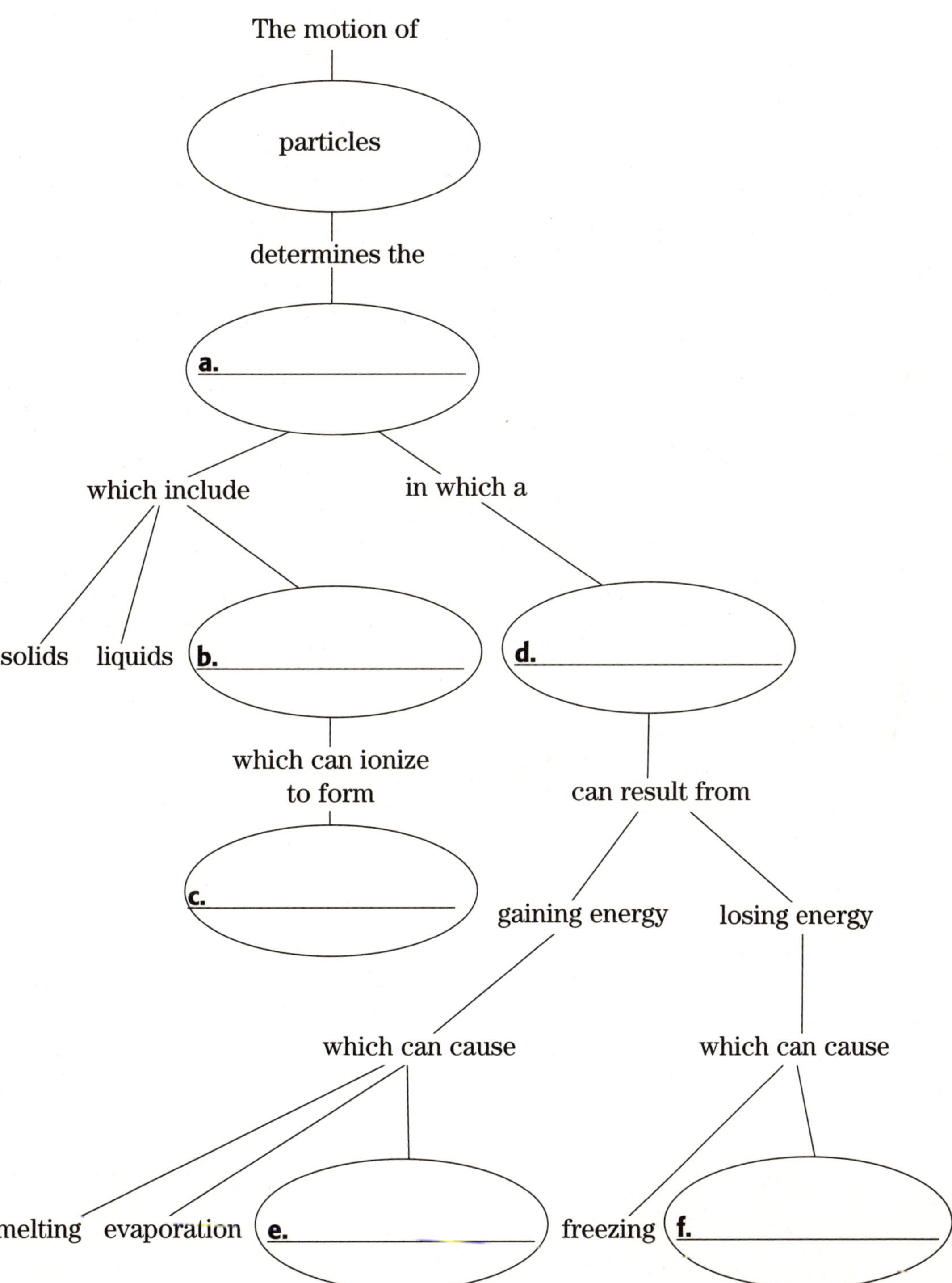

Assessment

Performance-Based Assessment

Teacher Notes

PURPOSE

Students will observe the effects of adding heat energy to a glass of ice water. Students will then use their knowledge of temperature and changes of state to answer questions.

Rebecca Ferguson
Northridge Middle School
North Richland Hills,
Texas

TIME REQUIRED

One 45-minute class period. Students will need 15 minutes at the activity station and 30 minutes to answer the analysis questions.

RATING

Easy ←——1——2——3——4——→ Hard

Teacher Prep–1
Student Set-Up–1
Concept Level–2
Clean Up–1

ADVANCE PREPARATION

Equip each activity station with the necessary materials.

SAFETY CAUTION

Spilled water is a slipping hazard. Wipe up spills immediately. Hot water is a burn hazard. Have students use caution when running hot water over the milk container.

TEACHING STRATEGIES

This activity works best in groups of 2–3 students. The consistent temperature of the ice water demonstrates that temperature does not change until the change of state is complete.

BACKGROUND INFORMATION

Energy that is added during a change of state is used to break attractions between the particles. So, the temperature does not change until the change of state is complete.

Performance-Based Assessment *continued*

Evaluation Strategies

Use the following rubric to help evaluate student performance.

Rubric for Assessment

Possible points	Appropriate use of materials and equipment (10 points possible)
10–7	Successful completion of activity; safe and careful handling of materials and equipment; attention to detail; superior lab skills
6–4	Activity is generally complete; successful use of materials and equipment; sound knowledge of lab techniques; somewhat unfocused performance
3–1	Attempts to complete activity yield inadequate results; sloppy lab technique; apparent lack of skill
	Quality and clarity of observations (50 points possible)
50–40	Superior observations stated clearly and accurately; high level of detail; correct usage of units of measurement
39–20	Accurate observations; moderate level of detail; correct usage of units of measurement
19–10	Complete observations but expressed in unclear manner; may include minor inaccuracies; attempts to use units of measurement include inconsistencies
9–1	Erroneous, incomplete, or unclear observations; lack of accuracy, details, units of measurement
	Analysis of observations (40 points possible)
40–25	Clear, detailed explanations show superior knowledge of temperature and changes of state
24–15	Adequate understanding of temperature and changes of state with minor difficulty in expression
14–1	Poor understanding of temperature and changes of state; explanations unclear or not relevant to activity; substantial factual errors

Performance-Based Assessment

OBJECTIVE

Observe the effects of adding heat energy to ice water.

KNOW THE SCORE!

As you work through the activity, keep in mind that you will be earning a grade for the following:

- how well you work with the materials and equipment (10%)
- the quality and clarity of your observations (50%)
- how well you analyze your observations (40%)

Using Scientific Methods

ASK A QUESTION

How is temperature affected during a change of state?

MATERIALS AND EQUIPMENT

- beaker or water glass, about 1/4 liter
- hot plate
- ice cubes
- thermometer
- water, cold

SAFETY INFORMATION 

- Wipe up spills immediately; spilled water is a slipping hazard.
- Use caution when using the hot plate; the hot plate is a burn hazard.

FORM A HYPOTHESIS

1. You will place a thermometer in a glass of ice water and place the glass on a hot plate. What do you think will happen to the temperature as the ice melts?

   ____

   ____

TEST THE HYPOTHESIS

2. Fill a glass or beaker about half way with cold tap water. Add several ice cubes, and insert a thermometer. Wait until the temperature stabilizes.

3. Place the glass on a hot plate set to low heat.

4. Observe and record the temperature of the water for three minutes or until the ice completely melts (whichever comes first).

Performance-Based Assessment *continued*

5. Keep the glass on the hot plate until the ice cubes melt completely. Observe and record what happens to the temperature of the water for three minutes after the last ice cube melts.

ANALYZE THE RESULTS

6. What happened to the ice cubes in step 3?

____

____

7. Explain what happened to the temperature of the water while the ice cubes melted.

____

____

____

8. Explain what happened to the temperature of the water after the ice cubes melted completely.

____

____

____

DRAW CONCLUSIONS

9. Why didn't the temperature of the water change in step 4?

____

____

____

10. Why did the temperature of the water rise in step 5?

____

____

____

Performance-Based Assessment *continued*

BIG IDEA QUESTION

11. Describe what happened to the particles in the ice cubes as the ice water was heated on the hot plate.

Standards Assessment

Teacher Notes and Answer Key

To provide practice under more realistic testing conditions, give students 20 min to answer all of the questions in this assessment.

QUESTION NUMBER	CORRECT ANSWER	STANDARD
1	*C*	8.3 (supporting)
2	*A*	8.3 (supporting)
3	*D*	8.9.d (supporting)
4	*A*	8.9.d (supporting)
5	*B*	8.3.d (mastering)
6	*B*	8.3.d (mastering)
7	*D*	8.3.d (mastering)
8	*B*	8.3.e (mastering)
9	*C*	8.3.e (mastering)
10	*D*	8.3.e (mastering)
11	*A*	8.3.d (mastering)
12	*A*	5.3.c (mastering)
13	*B*	5.6.a (mastering)
14	*D*	6.4.e (supporting)

TEST DOCTOR

The following Standards Assessment questions have been diagnosed by the Test Doctor. Find out what might be causing your students' "ailing" answers. Each Test Doctor is followed by a diagnostic teaching tip to help you address students' learning needs.

Question 1 *asks students to choose the correct meaning of the word* distinct *based on the context of a sentence.*

A **Incorrect.** The physical and chemical properties of an element might not be easy to see, hear, or smell, especially if the quantity of the element is made up of only a few atoms.

B **Incorrect.** Because an element can be as small as one atom, this meaning of *distinct* is not appropriate for this sentence.

C **Correct.** *Distinct* in this sentence means "clearly different and separate." Each element has a different atomic structure and is therefore distinct from all other elements.

D **Incorrect.** Because an element can be as small as one atom, this meaning of *distinct* is not appropriate for this sentence.

Standards Assessment *continued*

Diagnostic Teaching Tip: Students who have difficulty answering this type of question might benefit from exercises in which they must pick the correct sense of a word to fit a context. Have students explore the dictionary with the purpose of finding four words with multiple meanings. Ask students to write a sentence for each word and list the definitions below it. Have students exchange sentences with a partner and indicate which definitions they think fit the context of the sentences. Students can then discuss their choices with their partners.

Question 2 *asks students to identify a synonym for the word* structure.

A Correct. *Composition* and *structure* are synonyms. Both can refer to the way in which the parts of a whole are arranged.

B Incorrect. *Stability* does not mean the same thing as *structure*, although a structure might be described as stable.

C Incorrect. *Function* and *structure* are not the same thing, although certain structures in organisms often perform specific functions.

D Incorrect. *Combination* and *structure* do not mean the same thing, although structures might exist in different combinations.

Diagnostic Teaching Tip: Students who have difficulty answering this question correctly might benefit from practice using a thesaurus. Provide students with a paragraph of text with several words underlined. Ask students to use a thesaurus to find synonyms for the underlined words. Remind students that the sentences should still make sense when the original words have been replaced with synonyms.

Question 3 *asks students to choose the appropriate tense of the verb* to construct *for a sentence.*

A Incorrect. The past tense of the verb is not appropriate here.

B Incorrect. The future perfect tense of the verb is not appropriate here.

C Incorrect. The conditional mood is not appropriate here.

D Correct. Because the experiment is in the future, the graph is also in the future; therefore, the future tense is appropriate here.

Diagnostic Teaching Tip: Students who have difficulty answering this question correctly might benefit from conjugating several regular verbs in past, present, and future tenses. Help students create a chart for listing all the forms of regular verbs in past, present, and future tenses. Give them a list of verbs and ask them to fill in the chart.

Question 4 *asks students to define the word* interpret.

A Correct. *Interpret* means "to explain the meaning or significance of something."

B Incorrect. *Communicate* means "to give or exchange information."

C Incorrect. *Suggest* means "to offer something as a possible choice."

D Incorrect. *Describe* means "to give an account of details or characteristics."

Diagnostic Teaching Tip: Students who choose answer **B** may be confusing the broad idea of communication with the specific process of interpreting. Stress to students that a person can communicate an interpretation or interpret a communication but that they are not the same thing. Students who choose answer **C** may not feel comfortable interpreting observations or data, and so may shy away from the correct answer. Encourage students who choose answer **D** to differentiate between description, which goes hand in hand with observation, and interpretation, which involves analysis of those observations.

Question 5 *asks students to demonstrate understanding that the states of matter depend on molecular motion.*

A Incorrect. Freezing does affect the state of a substance, but the state of matter does not depend upon freezing.

B Correct. The movement of particles determines the state of matter of a given substance.

C Incorrect. The particles in a substance do not expand, although the distance between them will expand when the substance is heated.

D Incorrect. The particles in a substance do not shrink, although the distance between them will shrink when energy is removed from the substance.

Diagnostic Teaching Tip: Students who have difficulty answering this question correctly might benefit from drawing diagrams of the way molecules behave in a solid, a liquid, and a gas. Have students write captions to correspond with their diagrams, identifying the state of matter shown and describing the behavior of the molecules in each state.

Question 6 *asks students to demonstrate understanding that material changes caused by physical processes are reversible.*

A Incorrect. This statement is not true. A cooling substance moves directly from liquid to solid.

B Correct. If energy is added to this substance, it will return again to a liquid state.

C Incorrect. The substance in its liquid form could become a gas if enough energy is added to it, but it is now a solid.

D Incorrect. The substance will not remain solid if enough energy is applied to it to change its state. Physical changes are impermanent and reversible.

Diagnostic Teaching Tip: Students who answer this question incorrectly might benefit from conducting an experiment that illustrates the concept in question here. Give students a few crayons of the same color and ask them to record the crayons' physical properties. What color are they? Are they soft or hard? What purpose do they serve? Then, have students heat the crayons to their melting point and record their observations and the crayons' physical properties. Allow the crayons to cool again, and have the students examine them and record their physical properties. Discuss the results of the experiment. What happens if the crayons are reheated?

Standards Assessment *continued*

Question 7 *asks students to demonstrate ability to construct appropriate graphs from data and describe the relationships between variables.*

A Incorrect. The water molecules gain energy and move farther apart as the water changes state.

B Incorrect. The graph clearly shows energy being added.

C Incorrect. This is a false statement. Molecules become *less* ordered as the state changes from liquid to gas.

D Correct. The temperature will rise to 100°C, the boiling point of water at sea level, and then it will level off, remaining at 100°C until all the water has evaporated.

Diagnostic Teaching Tip: Students who struggle to answer this type of question might benefit from practice graphing the temperature of substances as they boil and freeze. Arrange for students to conduct experiments with two substances other than water, one that boils at a relatively low temperature, and one that freezes relatively easily. Have students prepare data collection tables for measuring the temperature of the substances at intervals of 30 seconds. When they have collected data for both substances, guide them in plotting their data on graphs showing the relationship between the temperature of the substance and the energy added or removed.

Question 8 *asks students to demonstrate knowledge of the different ways molecules move in solids, liquids, and gases.*

A Incorrect. The particles in a solid can only vibrate. Milk is a liquid.

B Correct. The particles in milk, a liquid, move fast enough to overcome some of the attractions between them, allowing them to slide past each other freely.

C Incorrect. The particles in plasma have broken apart; milk is obviously a liquid.

D Incorrect. The particles in a gas move about freely; milk is obviously a liquid.

Diagnostic Teaching Tip: Students who struggle to answer this type of question might benefit from an exercise classifying substances as solid, liquid, or gas. Brainstorm a list of common objects and substances, perhaps asking students to list the things they see in the classroom. Have students create a three-column graphic organizer and sort the items from your list into the appropriate places in the chart.

Question 9 *asks students to demonstrate understanding that the states of matter depend on molecular motion.*

A Incorrect. When energy is added to water, its molecules move more quickly, not more slowly.

B Incorrect. When energy is added to water, its molecules do move more quickly, but the temperature of the water rises.

C Correct. When liquid water becomes ice, it is because energy is removed from the water, causing its molecules to lock into place.

D Incorrect. When energy is removed from water, its molecules become more tightly connected rather than disconnected.

Diagnostic Teaching Tip: Students who have difficulty answering this question correctly might benefit from more practice observing the behavior of water as it boils and freezes. Have students place several ice cubes in a beaker and heat them, keeping close watch on the behavior of the water as it melts, boils, and evaporates. Ask students to write a short report matching their observations with what was happening to the water on a molecular level.

Question 10 *asks students to demonstrate understanding that physical processes change the form of matter without chemical reactions.*

A Incorrect. A substance loses energy when it freezes or condenses, but these are not the only ways the state of matter can change.

B Incorrect. A substance gains energy when it melts, evaporates, or sublimes, but these are not the only ways the state of matter can change.

C Incorrect. A change of state does not involve a chemical reaction. When a substance changes state, its physical properties change, but its chemical properties remain the same.

D Correct. A change of state is the change of a substance from one physical form to another.

Diagnostic Teaching Tip: Students who choose answers **A** or **B** might benefit from creating graphs to show the relationship between energy and temperature in freezing and melting. Students who choose answer **C** might benefit from creating a table that compares the chemical and physical properties of several common substances.

Question 11 *asks students to demonstrate understanding that substances can be classified by their properties.*

A Correct. Unlike gases, plasmas can conduct electrical current.

B Incorrect. Plasmas do not have a definite shape. Like gases, they take the shape of their containers.

C Incorrect. Plasmas do not have definite volume. Like the particles in gases, the particles in plasmas will spread apart if allowed to do so.

D Incorrect. Plasmas, unlike gases, are affected by magnetism. Sometimes magnetic fields are used to contain very hot plasmas that would destroy a solid container.

Diagnostic Teaching Tip: Students who have difficulty answering this question correctly might benefit from researching plasmas in the library or on the Internet. Make a class list of all the substances students have learned can be classified as plasmas. Have students share the most unusual facts they learned about plasmas with the class. Are there any examples of plasma in the classroom?

Question 12 *asks students to demonstrate understanding that changes in heat, air movement, and humidity result in changes of weather.*

A Correct. Precipitation occurs when moist air is cooled below its condensation point (dew point). This is the point at which water vapor, a gas, changes states and becomes liquid again.

B Incorrect. Precipitation does not occur at the point of evaporation.

C Incorrect. Precipitation does not occur at this point in the water cycle.

D Incorrect. Precipitation does not occur at this point in the water cycle.

Diagnostic Teaching Tip: Students who have difficulty answering this question correctly might benefit from reviewing the stages of the water cycle with changes in states in mind. Have students create a diagram of the water cycle, labeling the points at which water changes its state.

Question 13 *asks students to classify a substance according to defined criteria.*

A Incorrect. If the substance were a gas, its molecules would be broken apart from each other.

B Correct. The substance is a liquid at its boiling point, when the pressure inside the bubbles equals the atmospheric pressure.

C Incorrect. The molecules in a solid move only enough to vibrate against each other.

D Incorrect. The molecules in a liquid at its freezing point would be slowing down rather than moving quickly.

Diagnostic Teaching Tip: Students who have difficulty answering this question correctly might benefit from identifying other mystery substances using a set of clues. Have students choose a substance—either a specific substance or a solid, liquid, or gas at a particular point of transition—and write clues as to its identity. Have students exchange their clues and identify the mystery substances.

Question 14 *asks students to demonstrate understanding that Earth's atmosphere exerts pressure that decreases with distance above Earth's surface.*

A Incorrect. Atmospheric pressure changes from one altitude to another.

B Incorrect. Atmospheric pressure at sea level is the same at every location.

C Incorrect. This statement is the *opposite* of the true statement.

D Correct. As an object moves farther from Earth's surface, atmospheric pressure decreases because the mass of air above that object decreases.

Diagnostic Teaching Tip: Students who have difficulty answering this question correctly might benefit from reviewing how to make and interpret tables. Give the students a data set. Direct them to identify the dependent variables and the independent variable and then organize the data into data pairs. Next, have them organize the data pairs into a table. Remind them to label the columns and header.

Name _________________________________ Class _______________ Date ____________

Standards Assessment

REVIEWING ACADEMIC VOCABULARY

______ **1.** In the sentence "Each element has distinct physical and chemical properties," what does the word *distinct* mean?
   A easy to hear, see, or smell
   B large enough to be noticed
   C clearly different and separate
   D very great in degree

______ **2.** Which of the following words is the closest in meaning to the word *structure*?
   A composition
   B stability
   C function
   D combination

______ **3.** Which of the following sets of words best completes the following sentence: "After they conduct their experiment, the students ______ a graph based on their data"?
   A have constructed
   B will have constructed
   C would construct
   D will construct

______ **4.** Which of the following words means "to figure out the meaning of"?
   A interpret
   B communicate
   C suggest
   D describe

REVIEWING CONCEPTS

______ **5.** The state of matter of a substance depends upon how the particles in that substance
   A freeze.
   B move.
   C expand.
   D shrink.

______ **6.** A substance changes state from a liquid to a solid. Which of the following is true of that substance?
   A It passes through a plasma state.
   B It can return to a liquid state.
   C It will soon become a gas.
   D It will remain permanently solid.

▌Standards Assessments

Adding Energy to Water

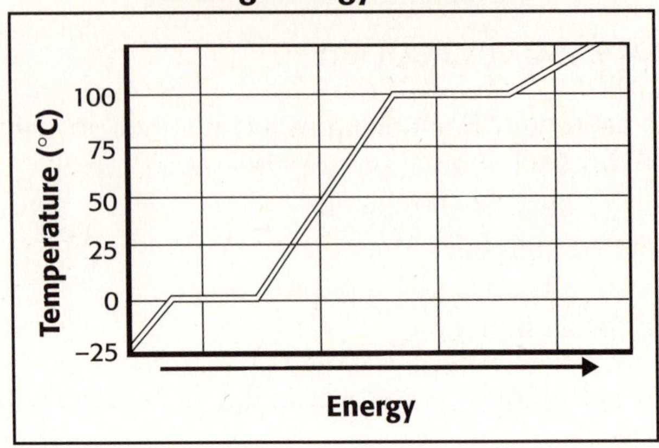

______ **7.** The graph above shows the effect of adding energy to water at sea level. When the temperature reaches 100°C, what happens to the water molecules as energy continues to be added?

   A The water molecules gain energy as the temperature continues to rise.

   B The water molecules gain no energy and the temperature stays the same.

   C The water molecules become more ordered as the state changes to a gas.

   D The water molecules move farther apart as the state changes to a gas.

______ **8.** Which of the following statements best describes the particles contained in a glass of milk?

   A They are closely locked into position and can only vibrate.

   B They are loosely connected and can slide past each other.

   C They have broken apart.

   D They move about freely.

______ **9.** Which of the following sentences best describes the process that occurs when liquid water becomes ice?

   A Energy is added to the water, so its molecules move more slowly.

   B Energy is added to the water, so its molecules move more quickly.

   C Energy is removed from the water, so its molecules lock into place.

   D Energy is removed from the water, so its molecules move apart.

______ **10.** A change in the state of matter always includes

   A a loss of energy.

   B a gain of energy.

   C a change in the chemical properties of a substance.

   D a change in the physical form of a substance.

Standards Assessment *continued*

_______ **11.** Plasma is the most common state of matter in the universe. How are
plasmas different from gases?
A Plasmas conduct electric currents.
B Plasmas have a definite shape.
C Plasmas have a definite volume.
D Plasmas are unaffected by magnetism.

REVIEWING PRIOR LEARNING

_______ **12.** Precipitation occurs at the point where
A moist air is cooled below its condensation point.
B dry air picks up water through evaporation.
C warm air rises into the atmosphere and cools.
D cool air sinks because of convection in the atmosphere.

_______ **13.** The molecules in an unknown substance collide with and slide past
each other. They are moving quickly enough that the substance's vapor
pressure equals the atmospheric pressure. How would you classify the
unknown substance?
A It is a gas at its condensation point.
B It is a liquid at its boiling point.
C It is a solid at its melting point.
D It is a liquid at its freezing point.

ALTITUDE AND ATMOSPHERIC PRESSURE

Altitude above sea level (m)	Atmospheric pressure (kPa)
0	101
100	98.2
1000	85.8
10000	20.4

_______ **14.** According to the table above, what is the relationship between
atmospheric pressure and altitude?
A Atmospheric pressure remains constant at any altitude.
B Atmospheric pressure differs at sea level according to location.
C Atmospheric pressure increases as altitude increases.
D Atmospheric pressure decreases as altitude increases.

DATASHEET

A Change of State

Teacher Notes

This activity has students witness and experience the results of a change of state (covers standards 8.3.d and 8.3.e). Only a small amount of alcohol is needed for this activity. Demonstrate how little alcohol is needed by pouring a sufficient amount into a cup.

MATERIALS

For each group

- cotton swab
- cup, plastic, small
- rubbing alcohol

SAFETY CAUTION

Remind students to review all safety cautions and icons before beginning this activity.

Explore Activity) **DATASHEET A**

A Change of State

In this activity, you will use rubbing alcohol (isopropyl alcohol) to investigate a change of state.

SAFETY INFORMATION 

PROCEDURE

1. Get **rubbing alcohol** and a **small plastic cup.**
  • Pour some alcohol into the cup until the alcohol just covers the bottom of the cup.

2. Dip the tip of a **cotton swab** into the alcohol in the cup.

3. Make sure that there are no cuts or scratches on your hand.
  • Rub the cotton swab on the palm of your hand.

4. Did you see, hear, feel, or smell anything?

____

____

____

5. Wash your hands thoroughly.

ANALYSIS

6. Explain what happened to the alcohol that you put on your hand. What change of state occurred?

____

____

____

7. Did you sense hot or cold?

____

  • If so, how can you explain it?

____

____

____

A Change of State *continued*

8. Think about the alcohol on your hand. How did the alcohol particles change? (Circle the letter of the correct answer.)

a. They spread out and changed to a gas state.

b. They moved closer together and changed into a solid state.

c. They spread out and changed into a plasma state.

d. They moved closer together and remained a liquid.

| Explore Activity | **DATASHEET B** |

A Change of State

In this activity, you will use rubbing alcohol (isopropyl alcohol) to investigate a change of state.

SAFETY INFORMATION 

PROCEDURE

1. Pour **rubbing alcohol** into a **small plastic cup** until the alcohol just covers the bottom of the cup.

2. Moisten the tip of a **cotton swab** by dipping it into the alcohol in the cup.

3. Rub the cotton swab on the palm of your hand. Make sure that there are no cuts or abrasions on your hand.

4. Record your observations.

____

____

5. Wash your hands thoroughly.

ANALYSIS

6. Explain what happened to the alcohol after you rubbed the swab on your hand.

____

____

7. Did you feel a sensation of hot or cold? If so, how do you explain what you observed?

____

____

8. Describe the change in the alcohol based on what happened to its particles.

____

____

____

A Change of State

DATASHEET C

In this activity, you will use rubbing alcohol (isopropyl alcohol) to investigate a change of state.

SAFETY INFORMATION 

PROCEDURE

1. Pour **rubbing alcohol** into a **small plastic cup** until the alcohol just covers the bottom of the cup.

2. Moisten the tip of a **cotton swab** by dipping it into the alcohol in the cup.

3. Rub the cotton swab on the palm of your hand. Make sure that there are no cuts or abrasions on your hand.

4. Record your observations.

____

____

5. Wash your hands thoroughly.

ANALYSIS

6. Explain what happened to the alcohol after you rubbed the swab on your hand.

____

____

7. Did you feel a sensation? If so, how do you explain what you observed?

____

____

____

8. Describe the change in the alcohol particles.

____

____

____

____

Quick Lab)

DATASHEET

Changing Volumes

Teacher Notes

This activity will allow students to observe how a gas and a liquid behave under compression (covers 8.3.d and 8.3.e).

MATERIALS

For each group

- syringe
- water

SAFETY CAUTION

Remind students to review all safety cautions and icons before beginning this activity.

Quick Lab

Changing Volumes

SAFETY INFORMATION

PROCEDURE

1. Draw 10 mL of air into a **syringe.**
   - Tighten the cap.

2. Push in the plunger.
   - Write your results.

   ____

   - Could you push the plunger?

   ____

   - Did you hear air coming out?

   ____

   - Did you see anything that changed?

   ____

3. Repeat steps 1 and 2 using **10 mL of water.**
   - Write your results.

   ____

   - Could you push the plunger?

   ____

   - Did you hear air coming out?

   ____

   - Did you see anything that changed?

   ____

   ____

   ____

Changing Volumes *continued*

4. Explain the difference between using air and using water.
   • Remember that gas particles have space between them and liquid particles are close together.

____

____

____

____

____

5. If you froze the water in the syringe from step 3 into ice, a solid, could this solid be squeezed down?
   • In your answer, talk about how close particles are in solids.

____

____

____

____

____

Quick Lab

DATASHEET B

Changing Volumes

SAFETY INFORMATION

PROCEDURE

1. Draw 10 mL of air into a **syringe.** Tighten the cap.

2. Push in the plunger. Record your results.

____

____

____

3. Repeat steps 1 and 2 using **10 mL of water.**

____

____

____

4. Explain any difference in your results in terms of the particles in each material.

____

____

____

____

____

5. Can a solid be compressed? Explain, in terms of the particles in the solid.

____

____

____

Name _______________________________ Class _______________ Date _______________

Changing Volumes

SAFETY INFORMATION

PROCEDURE

1. Draw 10 mL of air into a **syringe.** Tighten the cap.
2. Push in the plunger. Record your results.

____

____

____

3. Repeat steps 1 and 2 using **10 mL of water.**

____

____

____

4. Explain any difference in your results in terms of the characteristics of states of matter.

____

____

____

____

____

5. Can a solid be compressed? Explain, in terms of the state of matter.

____

____

____

____

Quick Lab

DATASHEET

Boiling Water Without Heating It

Teacher Notes

This activity investigates the connection between pressure and boiling point (covers standards 8.5.d and 8.7.c).

MATERIALS

For each group

- syringe
- warm water

SAFETY CAUTION

Remind students to review all safety cautions and icons before beginning this activity.

(Quick Lab)   **DATASHEET A**

Boiling Water Without Heating It

You have seen water boil when it is heated. However, air pressure above the water plays a role, too. In this activity, you will explore the connection between pressure and boiling point.

SAFETY INFORMATION 

TRY IT!

1. Remove the **cap** from a **syringe.**

2. Your teacher will give you some warm water.
   - Place the tip of the syringe in the **warm water.**
   - Pull the plunger out until you have 10 mL of water in the syringe.

3. Place the cap tightly on the syringe.
   - Hold the syringe.
   - Slowly pull the plunger out.

4. What happened to the water?

   ____

   ____

THINK ABOUT IT!

5. Put a check mark next to each of the following that may cause water to boil.

   ____ Air pressure gets higher.

   ____ Water gets hotter.

   ____ Air pressure gets lower.

   ____ More water is added.

6. Think about when you pulled the plunger out of the syringe. Did the air pressure in the syringe go up, go down, or stay the same?

   ____

7. Use these words to fill in the blanks: water, temperature, air pressure

   The low _____________________ caused the _____________________ to

   boil at a low _____________________.

8. Use one of these words to fill in the blank: temperature, air pressure

   Water usually boils because the temperature goes up. But, in this case, water

   boiled because the _____________________ changed.

| Quick Lab | **DATASHEET B** |

Boiling Water Without Heating It

You have seen water boil when it is heated. However, air pressure above the water plays a role, too. In this activity, you will explore the connection between pressure and boiling point.

SAFETY INFORMATION

TRY IT!

1. Remove the **cap** from a **syringe.**

2. Place the tip of the syringe in the **warm water** that is provided by your teacher. Pull the plunger out until you have 10 mL of water in the syringe.

3. Place the cap tightly on the syringe. Hold the syringe, and slowly pull the plunger out.

4. Record any changes you see in the water.

____

____

THINK ABOUT IT!

5. If the temperature stays the same, what must happen to the atmospheric pressure in order to cause the liquid to boil?

____

6. What happened to the air inside the syringe above the water as you pulled the plunger out of the syringe?

____

7. How did this cause the water to boil?

____

8. When you normally see water boil, what is different about the conditions under which that happens, compared with what you just saw?

____

____

Boiling Water Without Heating It

You have seen water boil when it is heated. However, air pressure above the water plays a role, too. In this activity, you will explore the connection between pressure and boiling point.

SAFETY INFORMATION

TRY IT!

1. Remove the **cap** from a **syringe.**

2. Place the tip of the syringe in the **warm water** that is provided by your teacher. Pull the plunger out until you have 10 mL of water in the syringe.

3. Place the cap tightly on the syringe. Hold the syringe, and slowly pull the plunger out.

4. Record any changes you see in the water.

____

____

THINK ABOUT IT!

5. What must happen to cause liquid to boil at a fixed temperature?

____

____

6. What happened to the air inside the syringe above the water as you pulled the plunger out of the syringe?

____

____

7. What caused the water to boil?

____

8. Compare your results to typical situations in which water boils.

____

____

____

Boiling and Temperature

Teacher Notes

This activity has students investigate the relationship between heating and temperature through various states of matter for water (covers standards 8.5.d and 8.7.c). They will construct graphs from their data and analyze them to find quantitative relationships between temperature and heating (covers standards 8.9.d and 8.9.e).

Anna Marcello
Audubon Middle School
Los Angeles, California

TIME REQUIRED

One 45-min class period

LAB RATINGS

Teacher Prep–3
Student Set-Up–2
Concept Level–3
Clean Up–2

Easy ←——— 1　　2　　3　　4 ———→ Hard

MATERIALS

The materials listed on this page are enough for a group of three to four students.

SAFETY CAUTION

Remind students to review all safety cautions and icons before beginning this activity.

PREPARATION NOTES

To construct the wire-loop stirring device, make a small loop at one end of a 25-cm piece of copper wire. The loop should fit easily into the graduated cylinder with the thermometer in place. Angle the loop so that it is perpendicular to the rest of the wire. At the other end of the wire, make a handle that extends in the opposite direction of the loop. Place the loop around the thermometer, and use the handle to move the device up and down.

DATASHEET A

Boiling and Temperature

When you add energy to a substance through heating, does the substance's temperature always go up? When you remove energy from a substance through cooling, does the substance's temperature always go down? In this lab, you'll investigate these important questions with a very common substance—water.

OBJECTIVE

Measure and record time and temperature accurately.

Graph the temperature change of water as it changes state.

Analyze and interpret graphs of changes of state.

MATERIALS

- beaker, 250 mL or 400 mL
- coffee can, large
- gloves, heat-resistant
- graduated cylinder, 100 mL
- graph paper
- hot plate
- ice, crushed
- rock salt
- stopwatch
- thermometer
- water
- wire-loop stirring device

SAFETY INFORMATION

PROCEDURE

1. Fill the beaker about one-third to one-half full with water.

2. Put on hot-pad gloves.
- Turn on the hot plate.
- Put the beaker on the hot plate.
- Put the thermometer in the beaker.
- **Caution:** DO NOT touch the hot plate.

3. Record the temperature of the water every 30 seconds in the data table on the next page.
- Continue doing this until about one-fourth of the water boils away.
- Watch for the water to come to a full boil.
- Write the temperature and mark it "full boil."

Boiling and Temperature *continued*

Data Table								
Time (s)	30	60	90	120	150	180	210	240
Temperature (°C)								
Time (s)	270	300	330	360	390	420	450	480
Temperature (°C)								

4. Turn off the hot plate.

5. Let the beaker cool.
- Make a graph with temperature on the side (*y*-axis) and the time on the bottom (*x*-axis).
- Draw an arrow pointing to the "full boil" temperature.

6. After you finish the graph, use hot-pad gloves to pick up the beaker.
- Pour the warm water out.
- Rinse the warm beaker with cool water.
- **Caution:** Leave the gloves on. Even after cooling, the beaker is still too warm to handle without gloves.

7. Put about 20 mL of water in the graduated cylinder. (Hint: Make sure to look at eye level when you read the volume.)

8. Put the graduated cylinder in the coffee can.
- Fill in around the graduated cylinder with crushed ice.
- Pour rock salt on the ice around the graduated cylinder.
- Place the thermometer and the wire-loop stirring tool in the graduated cylinder.

9. The ice will melt and mix with the rock salt. The level of ice will decrease.
- Add ice and rock salt to the can, as needed, to keep it full.

10. Record the temperature of the water in the graduated cylinder every 30 seconds in the data table below.
- Stir the water with the stirring tool.
- **Caution:** DO NOT stir the water with the thermometer.

11. Once the water begins to freeze, stop stirring.
- **Caution:** DO NOT try to pull the thermometer out of the solid ice in the cylinder.

Data Table								
Time (s)	30	60	90	120	150	180	210	240
Temperature (°C)								
Time (s)	270	300	330	360	390	420	450	480
Temperature (°C)								

12. Watch for ice to start forming in the water. Write the temperature, and mark it "ice started."
 • Continue taking readings until the water in the graduated cylinder is all frozen.

13. Make another graph with temperature on the side (y-axis) and time on the bottom (x-axis).
 • Draw an arrow pointing to the "ice started" temperature.

ANALYZE THE RESULTS

14. Describing Events When you heat water, you add energy to it. What happens to the temperature of boiling water when you add energy to it?

____

____

15. Describing Events When you cool water, you remove energy from it. What happens to the temperature of freezing water when you remove energy from it?

____

____

16. Analyzing Data The slopes on the graphs show the rate

of ____________________ change.

17. Analyzing Results The hot-water graph has two slopes. The first one shows temperatures before the water started boiling. The second one shows temperatures after the water started boiling.

 a. Tell how the slopes compare in how they look: The first slope is

____________________ and the second slope is ____________________.

 b. Tell why these two slopes are different: In the first slope, adding energy

makes the water ____________________. In the second slope, adding

energy makes ____________________ instead of making the water hotter.

Boiling and Temperature *continued*

18. Analyzing Results The cold-water graph has two slopes. The first one shows temperatures before the water started freezing. The second one shows temperatures after the water started freezing.

a. Tell how the slopes compare in how they look: The first slope is

_______________________ and the second slope is _______________________.

b. Tell why these two slopes are different: In the first slope, removing energy

makes the water _______________________. In the second slope, removing

energy makes _______________________ instead of making the water colder.

DRAW CONCLUSIONS

19. Draw Conclusions What happens to temperature during a change of state (as when water is boiling or freezing)?
* Does the temperature go up?

____

* Does the temperature go down?

____

* Does the temperature stay the same?

____

BIG IDEA QUESTION

20. Applying Concepts The particles that make up solids, liquids, and gases are always moving. Adding or removing energy causes changes in this movement.

a. Look at the hot-water graph.
* Does the part with the slanted slope show that the water molecules are slowing down, speeding up, or not moving?

____

* Does the part with the flat line show that the water molecules are slowing down, changing state, or disappearing?

____

b. Look at the cold-water graph.
* Does the part with the slanted slope show that the water molecules are slowing down, speeding up, or not moving?

____

* Does the part with the flat line show that the water molecules are changing state, speeding up, or disappearing?

____

Boiling and Temperature

When you add energy to a substance through heating, does the substance's temperature always go up? When you remove energy from a substance through cooling, does the substance's temperature always go down? In this lab, you'll investigate these important questions with a very common substance—water.

OBJECTIVES

Measure and record time and temperature accurately.

Graph the temperature change of water as it changes state.

Analyze and interpret graphs of changes of state.

MATERIALS

- beaker, 250 mL or 400 mL
- coffee can, large
- gloves, heat-resistant
- graduated cylinder, 100 mL
- graph paper
- hot plate
- ice, crushed
- rock salt
- stopwatch
- thermometer
- water
- wire-loop stirring device

SAFETY INFORMATION 

PROCEDURE

1. Fill the beaker about one-third to one-half full with water.

2. Put on heat-resistant gloves. Turn on the hot plate, and put the beaker on it. Put the thermometer in the beaker. **Caution:** Be careful not to touch the hot plate.

3. Use the Data table below. Record the temperature of the water every 30 s. Continue doing this until about one-fourth of the water boils away. Note the first temperature reading at which the water is steadily boiling.

Data Table								
Time (s)	30	60	90	120	150	180	210	240
Temperature (°C)								
Time (s)	270	300	330	360	390	420	450	480
Temperature (°C)								

Boiling and Temperature *continued*

4. Turn off the hot plate.

5. While the beaker is cooling, make a graph of temperature (*y*-axis) versus time (*x*-axis) in the space below. Draw an arrow pointing to the first temperature at which the water was steadily boiling.

6. After you finish the graph, use heat-resistant gloves to pick up the beaker. Pour the warm water out, and rinse the warm beaker with cool water. **Caution:** Even after cooling, the beaker is still too warm to handle without gloves.

7. Put approximately 20 mL of water in the graduated cylinder.

8. Put the graduated cylinder in the coffee can, and fill in around the graduated cylinder with crushed ice. Pour rock salt on the ice around the graduated cylinder. Place the thermometer and the wire-loop stirring device in the graduated cylinder.

9. As the ice melts and mixes with the rock salt, the level of ice will decrease. Add ice and rock salt to the can as needed.

10. Use the Data table below. Record the temperature of the water in the graduated cylinder every 30 s. Stir the water with the stirring device. **Caution:** Do not stir with the thermometer.

Data Table								
Time (s)	30	60	90	120	150	180	210	240
Temperature (°C)								
Time (s)	270	300	330	360	390	420	450	480
Temperature (°C)								

11. Once the water begins to freeze, stop stirring. **Caution:** Do not try to pull the thermometer out of the solid ice in the cylinder.

12. Note the temperature when you first notice ice crystals forming in the water. Continue taking readings until the water in the graduated cylinder is completely frozen.

Boiling and Temperature *continued*

13. Make a graph of temperature (*y*-axis) versus time (*x*-axis) in the space below. Draw an arrow to the temperature reading at which the first ice crystals form.

ANALYZE THE RESULTS

14. Describing Events What happens to the temperature of boiling water when you continue to add energy through heating?

____

____

15. Describing Events What happens to the temperature of freezing water when you continue to remove energy through cooling?

____

____

6. Analyzing Data What does the slope of each graph represent?

____

____

17. Analyzing Results How does the slope of the graph that shows water boiling compare with the slope of the graph before the water starts to boil? Why is the slope different for the two periods?

____

____

____

8. Analyzing Results How does the slope of the graph showing water freezing compare with the slope of the graph before the water starts to freeze? Why is the slope different for the two periods?

____

____

____

DRAW CONCLUSIONS

19. Drawing Conclusions Using your answers to the previous two questions, make
a general statement that describes what happens to temperature during a
change of state.

____

____

____

____

____

____

BIG IDEA QUESTION

20. Applying Concepts The particles that make up solids, liquids, and gases are
in constant motion. Adding or removing energy causes changes in the movement of these particles. Using this idea, describe in terms of the motion of
the water molecules what is happening at each time represented in your two
graphs.

____

____

____

____

____

____

____

____

____

____

____

Name _________________________________ Class _______________ Date ____________

Boiling and Temperature

When you add energy to a substance through heating, does the substance's temperature always go up? When you remove energy from a substance through cooling, does the substance's temperature always go down? In this lab, you'll investigate these important questions with a very common substance—water.

OBJECTIVE

Measure and record time and temperature accurately.

Graph the temperature change of water as it changes state.

Analyze and interpret graphs of changes of state.

MATERIALS

- beaker, 250 mL or 400 mL
- coffee can, large
- gloves, heat-resistant
- graduated cylinder, 100 mL
- graph paper
- hot plate
- ice, crushed
- rock salt
- stopwatch
- thermometer
- water
- wire-loop stirring device

SAFETY INFORMATION 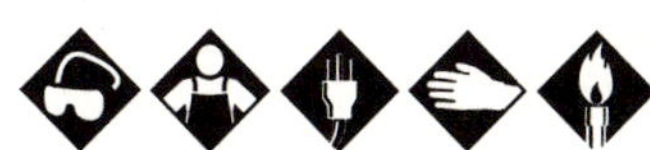

PROCEDURE

1. Fill the beaker about one-third to one-half full with water.

2. Put on heat-resistant gloves. Turn on the hot plate, and put the beaker on it. Put the thermometer in the beaker. **Caution:** Be careful not to touch the hot plate.

3. Record the temperature of the water every 30 s in the data table below. Continue doing this until about one-fourth of the water boils away. Note the first temperature reading at which the water is steadily boiling.

Data Table								
Time (s)	30	60	90	120	150	180	210	240
Temperature (°C)								
Time (s)	270	300	330	360	390	420	450	480
Temperature (°C)								

4. Turn off the hot plate.

Boiling and Temperature *continued*

5. While the beaker is cooling, make a graph of temperature (y-axis) versus time (x-axis). Draw an arrow pointing to the first temperature at which the water was steadily boiling.

6. After you finish the graph, use heat-resistant gloves to pick up the beaker. Pour the warm water out, and rinse the warm beaker with cool water. **Caution:** Even after cooling, the beaker is still too warm to handle without gloves.

7. Put approximately 20 mL of water in the graduated cylinder.

8. Put the graduated cylinder in the coffee can, and fill in around the graduated cylinder with crushed ice. Pour rock salt on the ice around the graduated cylinder. Place the thermometer and the wire-loop stirring device in the graduated cylinder.

9. As the ice melts and mixes with the rock salt, the level of ice will decrease. Add ice and rock salt to the can as needed.

10. Record the temperature of the water in the graduated cylinder every 30 s in the data table below. Stir the water with the stirring device. **Caution:** Do not stir with the thermometer.

Data Table								
Time (s)	30	60	90	120	150	180	210	240
Temperature (°C)								
Time (s)	270	300	330	360	390	420	450	480
Temperature (°C)								

11. Once the water begins to freeze, stop stirring. **Caution:** Do not try to pull the thermometer out of the solid ice in the cylinder.

12. Note the temperature when you first notice ice crystals forming in the water. Continue taking readings until the water in the graduated cylinder is completely frozen.

13. Make a graph of temperature (y-axis) versus time (x-axis). Draw an arrow to the temperature reading at which the first ice crystals form.

ANALYZE THE RESULTS

14. Describing Events What happens to the temperature of boiling water when you continue to add energy through heating?

____

15. Describing Events What happens to the temperature of freezing water when you continue to remove energy through cooling?

____

Boiling and Temperature *continued*

16. Analyzing Data What does the slope of each graph represent?

____

____

17. Analyzing Results Explain the slope changes in different parts of the
hot-water graph.

____

____

____

18. Analyzing Results Explain the slope changes in different parts of the
cold-water graph.

____

____

____

DRAW CONCLUSIONS

19. Drawing Conclusions Using your answers to the previous two questions, make
a general statement that describes what happens to temperature during a
change of state.

____

____

____

BIG IDEA QUESTION

20. Applying Concepts Explain in terms of the motion of the water molecules
what is happening at each time represented in your two graphs.

____

____

____

____

____

Science Skills Activity

Linear and Nonlinear Relationships

Teacher Notes

This activity helps students learn the difference between linear and nonlinear relationships on a graph (covers standard 8.9.g).

MATERIALS

- graph paper
- pencil
- ruler

Linear and Nonlinear Relationships

INVESTIGATION AND EXPERIMENTATION

8.9.g Distinguish between linear and nonlinear relationships on a graph of data.

TUTORIAL

1. The graph below represents an example of a linear relationship. You can tell that the relationship is linear because when you graph the two factors, they will form a straight line.

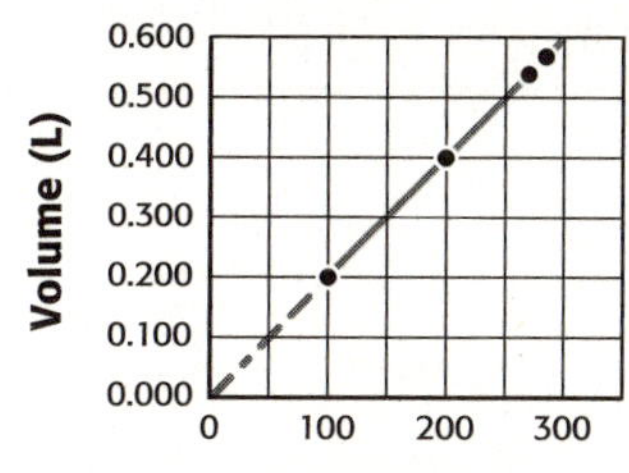

2. You can also tell that a relationship is linear by analyzing the data set that represents the relationship. For a relationship that is linear, as in the above graph, choose any two pairs of data. Then, divide each pair the same way, and you will get the same number, as shown below.

$$\frac{0.200 \text{ L}}{100 \text{ K}} = 0.002 \text{ L/K} \qquad\qquad \frac{0.400 \text{ L}}{200 \text{ K}} = 0.002 \text{ L/K}$$

3. A linear relationship is one that involves a constant rate of change. This was shown in the previous step because the two data pairs have the same ratio.

4. The graph below represents an example of a nonlinear relationship. A nonlinear relationship is one in which the rate of change over time between the two factors is not constant.

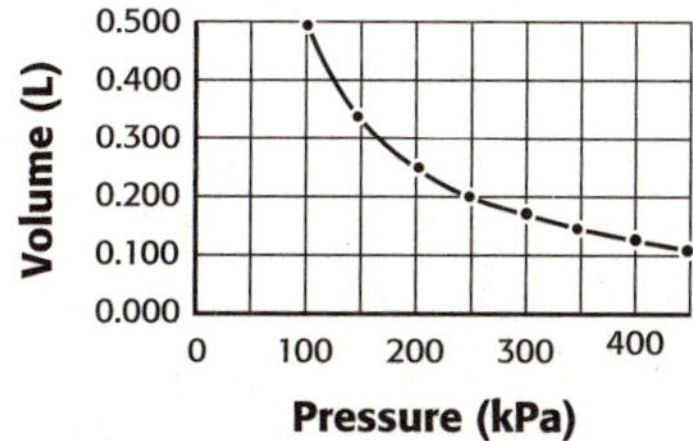

5. You can tell that a relationship between two factors is nonlinear by looking at its graph: it does not form a straight line. You can also tell that a relationship is nonlinear by doing the test described in item 2 above: two pairs of data will not have the same ratio.

Linear and Nonlinear Relationships *continued*

YOU TRY IT!

1. **Identifying Relationships** Analyze the data in the table below as was demonstrated in item 2 on the previous page. Is the relationship linear or nonlinear?

Pressure Vs. Temperature for a Gas at Constant Volume	
Pressure (atm)	**Temperature (K)**
1.0	100
2.0	200
3.5	350
4.0	400
5.0	500

____

____

____

2. **Evaluating Conclusions** Make a graph of the data from the table above. Did your graph confirm your initial conclusions? Explain.

____

____

____

____

3. **Analyzing Relationships** What does your analysis tell you about how the pressure and temperature of a gas vary together?

____

____

____

Answer Key

Directed Reading A

SECTION: FOUR STATES OF MATTER

1. B
2. A
3. A
4. C
5. B
6. A
7. C
8. C
9. A
10. D
11. A
12. A
13. D
14. C

SECTION: CHANGES OF STATE

1. D
2. C
3. A
4. C
5. D
6. B
7. B
8. A
9. C
10. D
11. A
12. D
13. A
14. B
15. C
16. D
17. C
18. D
19. A
20. B

Directed Reading B

SECTION: FOUR STATES OF MATTER

1. A state of matter is a physical form in which a substance can exist.
2. The three most familiar states of matter are solid, liquid, and gas.
3. atoms, molecules
4. A
5. C

6. B
7. C
8. A solid is the state of matter that has a definite shape and volume.
9. The particles in the liquid move quickly and slide past each other until the liquid takes the shape of the glass.
10. It shows that even when liquids change shape, they don't change volume.
11. A gas is a state of matter that has no definite shape or volume.
12. The tank contains helium particles that are forced close together. As helium enters the balloons, the atoms spread out, and the amount of empty space in the gas increases.
13. plasma
14. Plasmas conduct electric current, while gases do not. Electric and magnetic fields affect plasmas but do not affect gases.
15. Answers may vary. Sample answer: natural plasma: lighting; artificial plasma: fluorescent lights

SECTION: CHANGES OF STATE

1. A
2. change of state
3. melting, freezing, evaporation, condensation, sublimation
4. No, gallium's melting point is lower than your body temperature. It would melt in your hand.
5. melting point
6. freezing point
7. If energy is added, melting occurs. If energy is removed, freezing occurs.
8. B
9. C
10. A
11. atmospheric pressure, boiling point
12. condensation
13. boiling point
14. clump together
15. Solid carbon dioxide is called "dry ice" because instead of melting, it changes from a solid directly into a gas through sublimation.

16. sublimation
17. temperature
18. change of state

Vocabulary and Section Summary A

SECTION: FOUR STATES OF MATTER

1. states of matter: the physical forms of matter, which include solid, liquid, and gas
2. solid: the state of matter in which the volume and shape of a substance are fixed
3. liquid: the state of matter that has a definite volume but not a definite shape
4. gas: a form of matter that does not have a definite volume or shape
5. plasma: in physical science, a state of matter that starts as a gas and then becomes ionized; it consists of free-moving ions and electrons, it takes on an electric charge, and its properties differ from the properties of a solid, liquid, or gas

SECTION: CHANGES OF STATE

1. change of state: the change of a substance from one physical state to another
2. melting: the change of state in which a solid becomes a liquid by adding heat
3. evaporation: the change of state from a liquid to a gas
4. boiling: the conversion of a liquid to a vapor when the vapor pressure of the liquid equals the atmospheric pressure
5. condensation: the change of state from a gas to a liquid
6. sublimation: the process in which a solid changes directly into a gas

Vocabulary and Section Summary B

SECTION: FOUR STATES OF MATTER

1. plasma
2. solids
3. gases
4. liquids
5. states of matter
6. the way its particles interact

SECTION: CHANGES OF STATE

1. melting
2. condensation
3. sublimation
4. boiling
5. evaporation
6. change of state

G	D	I	T	H	P	I	J	B	H	Y	P	N	A	X
G	K	G	T	O	A	W	P	A	O	K	O	X	B	S
N	R	W	I	S	E	B	J	L	S	I	K	J	F	N
O	P	Y	U	U	N	T	O	P	T	L	L	N	O	W
I	Y	A	F	Z	D	U	Y	A	V	B	Z	I	Y	W
T	O	I	S	O	N	I	M	Y	L	S	T	J	N	L
A	P	A	Z	G	V	I	V	X	G	A	G	Y	D	G
R	K	U	Y	K	L	U	U	A	S	F	U	K	Q	I
O	I	N	S	B	F	T	V	N	Y	J	M	Y	S	K
P	H	W	U	H	D	N	E	L	H	D	E	I	D	S
A	Z	S	B	P	B	D	K	W	O	P	L	N	Z	I
V	H	C	H	A	N	G	E	O	F	S	T	A	T	E
E	J	I	Y	O	Z	U	I	B	J	Z	I	Y	B	W
E	P	X	C	P	S	T	S	O	H	K	N	W	O	X
A	O	U	W	I	J	U	I	A	U	O	G	T	Y	T

Reinforcement

MAKE A STATE-MENT

Liquid: Particles are close together; changes shape when placed in a different container; does not change in volume

Gas: Particles break away completely from one another; changes shape when placed in a different container; amount of empty space can change; changes volume to fill its container

Solid: Particles are close together; particles vibrate in place; particles are held tightly in place by other particles; has definite shape; does not change in volume

Critical Thinking

1. Answers may vary. Sample answer: It may be necessary to carry oxygen in portable containers because oxygen will probably not surround the planet evenly.
2. Answers may vary. Sample answer: Yes, it would be possible to make a fire, but only for a short time. In areas where oxygen is not present, oxygen may have to be supplied to the fire manually. Also, the wood will sublime at high temperatures, leaving no fuel for the fire.

3. Answers may vary. Sample answer: It would not be safe to visit without special protective clothing. It would be impossible to maintain liquids inside the human body because it is not a pressurized container.

4. Answers may vary. Sample answer: Phazon's cars would need pressurized gas tanks to keep gasoline inside. Their tires could be filled with a liquid instead of air.

SciLinks Activity

1. Solids: Particles are tightly packed; Particles do not move from place to place; keep their shape; do not flow easily

2. Liquids: Particles slide past each other; have a fixed volume, but not a fixed shape

3. Gases: Particles move freely; Particles move very fast; Particles are far apart; Volume changes; can be compressed
Venn diagram: Answers may vary. Sample answer: Solids, Liquids, and Gases: Particles vibrate; made up of atoms and molecules
Solids and Liquids: Particles are very close together; keep a fixed volume; not easily compressed
Liquids and Gases: Particles have no regular arrangement; assume the shape of the container; flow easily

Section Review

SECTION: FOUR STATES OF MATTER

1. Sample answer: Gas is a state of matter in which particles are far apart and moving quickly. Plasma is similar to gas, except its particles are broken apart and very hot.

2. Particles of a solid can only vibrate, particles of a liquid can slide past each other, and particles of a gas move quickly in any direction.

3. The state of matter shown in the jar is solid.

4. The particles cannot move around much relative to each other, but they can vibrate.

5. The volume of a gas can change because its particles are moving quickly and are far apart, so the space between its particles can be decreased or increased. The volume of a solid cannot change because its particles are right next to each other and cannot be squeezed together any further.

6. Tommy has not taken into account the fact that interactions between the particles will also determine the state of matter, and therefore particle motion. Each substance will have a characteristic interaction between its particles that will cause them to differ in particle motion under the same conditions.

SECTION: CHANGES OF STATE

1. Sample answer: Boiling is the change of state of a liquid to a gas at that liquid's boiling temperature, while melting is the change of state from a solid to a liquid.

2. Sample answer: Condensation is the change of state from a gas to a liquid, and evaporation is the opposite change of state.

3. As a substance freezes, the particles of the substance slow down and lock into place.

4. Boiling and evaporating both involve the change of state of a liquid to a gas. Boiling occurs when the temperature is high enough that all of a liquid's particles tend to vaporize, while evaporating happens at the surface of a liquid at any temperature.

5. If the liquid was not at its boiling point, the air pressure around the liquid must have decreased. If the liquid was at its boiling point, it must have been heated.

6. When the solid absorbs enough energy to melt and become a liquid, the particles can slide past one another and the material can flow.

7. The substance could be heated or the pressure of the air above the liquid could be lowered to boil the liquid.

8. 18 mL $\times$ 1,000 = 18,000 mL or 18 L

Chapter Review

1. B
2. Solid is the state of matter in which the substance has a definite shape and volume, and the particles are locked into place next to each other. Liquid is the state in which the substance has a definite volume, but takes the shape of its container.
3. Evaporation is the change of state of a liquid to a gas at the surface of a liquid. Boiling is the change of state of a liquid to a gas throughout a liquid.
4. Condensation is the change of state of a gas to a liquid. Sublimation is the change of state of a solid directly into a gas.
5. B
6. A
7. B
8. D
9. gases, liquids, solids
10. 100°C
11. The particles of liquid water move fast enough to be able to slide past each other, so a liquid will flow into the space available to it. An ice cube's particles are locked in place right next to each other, so they do not move relative to each other.
12. Through each state of matter, water's chemical identity remains the same. This is because changes of state are physical changes rather than chemical changes.
13. The boiling point of the substance is 80°C. The melting point of the substance is 20°C.
14. The temperature of the liquid will rise.
15. Answers should correctly describe the particles of the substance losing energy and moving less rapidly through the various state changes.
16. An answer to this exercise can be found at the end of the Teacher Edition.
17. The gas particles are moving rapidly and therefore pushing the tubes around as the gas particles are expelled from the tubes.
18. The temperature of the water must be 100°C.
19. Shower water vaporizes because of its high temperature and then condenses against the bathroom mirror. The mirror has a cool surface and absorbs energy from the higher-temperature water-vapor molecules.
20. Water has a stronger force of attraction between its particles because it takes more energy for the particles to separate from one another and change into a gas.
21. No; a change of state is a physical change that does not result in the chemical identity of the substance changing. The change described is a change of identity in the substance.
22. She could test the hypothesis by finding out if the gas bubbling up is water vapor.
23. The graph is linear. This tells you that there is a direct relationship between the surface area of a liquid and the amount that will evaporate in a given time.
24. Spraying water onto oranges protects the oranges from freezing because heat is given off by the water while it freezes, which increases the temperature of the oranges so that they do not freeze.

Section Quizzes

SECTION: FOUR STATES OF MATTER

1. D
2. B
3. C
4. E
5. D
6. A
7. B
8. C

SECTION: CHANGES OF STATE

1. B
2. A
3. F
4. C
5. D
6. E
7. H
8. G
9. I

Chapter Test A

1. B
2. D
3. B
4. C
5. C
6. D
7. D
8. A
9. B
10. B
11. A
12. C
13. D
14. C
15. B
16. A
17. solid
18. plasma
19. liquid
20. gas

Chapter Test B

1. B
2. D
3. B
4. C
5. C
6. A
7. D
8. A
9. B
10. D
11. F
12. E
13. A
14. C
15. D
16. B
17. C
18. B
19. C
20. D

Chapter Test C

1. condensation
2. evaporation
3. gas
4. state of matter
5. liquid
6. plasma

7. sublimation
8. melting
9. C
10. B
11. A
12. C
13. The temperature of the water remains constant (at the boiling point).
14. If energy is added, the ice will melt; if energy is removed, the water will freeze.
15. The boiling point of liquids decreases as atmospheric pressure decreases. At higher elevations, atmospheric pressure is reduced, so the boiling point of water will be lower than at sea level. This means that pasta boiled in Denver will cook at a lower temperature than pasta boiled in New Orleans. Therefore, the pasta will have to cook longer in Denver.
16. Dry ice changes directly from a solid to a gas, producing a smoke-like effect. Pouring water over the dry ice causes the carbon dioxide to sublime quickly. As a result, it turns to gas more quickly, producing a more dramatic smoke-like effect.
17. The colder weather is, the more helium will need to be pumped into the balloons for them to inflate to the desired level. This is because the colder the temperature, the slower the particles of gas move and the slower they hit the inside of the balloon. Thus, the colder it is, the lower the pressure. So, if it is cold out, more particles of gas will be needed in order to have the same level of pressure as with fewer particles on a warmer day.
18. A, gas; C, liquid; E, solid
19. B, condensation; D, freezing
20. **a.** states of matter, **b.** gases, **c.** plasma, **d.** change of state, **e.** sublimation, **f.** condensation

Performance-Based Assessment

1. Answers may vary. Sample answer: I think the temperature of the ice water will rise.
6. Answers may vary. Sample answer: The ice cubes began to melt.
7. Answers may vary. Sample answer: The temperature remained constant while the ice cubes melted.
8. Answers may vary. Sample answer: The temperature began to rise after the last ice cube had melted.
9. Answers may vary. Sample answer: The energy that was added during the change of state went to breaking the attractions between the particles that made up the ice rather than to changing the temperature of the ice water.
10. Answers may vary. Sample answer: Once all the attractions between the particles in the ice had been broken and the change of state from ice to liquid was complete, the heat energy from the hot plate could raise the temperature of the water.
11. Answers may vary. Sample answer: The particles that made up the solid ice absorbed heat from the hot plate and gained enough energy to overcome some of their attractions to each other to change state from solid to liquid.

Explore Activity

DATASHEET A
6. Sample answer: The alcohol disappeared by evaporating.
7. Sample answer: I felt a cool sensation. This happened because the alcohol absorbed heat from my hand.
8. A

DATASHEET B
6. The alcohol disappeared by evaporating.
7. Sample answer: I felt a cool sensation. This happened because the alcohol absorbed heat from my hand.
8. Sample answer: The alcohol particles became farther apart as they gained energy, and changed into the gas state.

DATASHEET C
6. The alcohol disappeared by evaporating.
7. Sample answer: I felt a cool sensation. This happened because the alcohol absorbed heat from my hand.
8. Sample answer: The alcohol particles became farther apart as they gained energy, and changed into the gas state.

Quick Lab: Changing Volumes

DATASHEET A
4. The gas could be compressed with effort because there is space between a gas's particles. But there is resistance to compression because a gas's particles move rapidly and collide with one another. The liquid could not be compressed because the liquid's particles are already very close together.
5. A solid could not be compressed because, like a liquid, its particles are already very close together.

DATASHEET B
4. The gas could be compressed with effort because there is space between a gas's particles. But there is resistance to compression because a gas's particles move rapidly and collide with one another. The liquid could not be compressed because the liquid's particles are already very close together.
5. A solid could not be compressed because, like a liquid, its particles are already very close together.

DATASHEET C
4. The gas could be compressed with effort because there is space between a gas's particles. But there is resistance to compression because a gas's particles move rapidly and collide with one another. The liquid could not be compressed because the liquid's particles are already very close together.
5. A solid could not be compressed because, like a liquid, its particles are already very close together.

Quick Lab: Boiling Water Without Heating It

DATASHEET A

5. Water gets hotter; air pressure gets lower.

6. The air pressure went down.

7. air pressure, water, temperature

8. air pressure

DATASHEET B

5. The pressure must decrease.

6. The air pressure inside the syringe decreased when the plunger was pulled out.

7. The air pressure became low enough for the water to boil at this temperature.

8. Normally, I see water boil because of heating rather than a change in pressure.

DATASHEET C

5. The pressure must decrease for a liquid to boil at a fixed temperature.

6. The air pressure inside the syringe decreased when the plunger was pulled out.

7. The air pressure became low enough for the water to boil at this temperature.

8. Sample answer: Typically, water boils because of heating rather than a change in pressure.

Chapter Lab

DATASHEET A

14. The temperature remains constant.

15. The temperature remains constant.

16. temperature

17. a. upward slanted, horizontal
b. hotter, steam

18. a. downward slanted, horizontal
b. colder, ice

19. no; no; yes, the temperature stays the same.

20. a. speeding up; changing state
b. slowing down; changing state

DATASHEET B

14. The temperature remains constant.

15. The temperature remains constant.

16. The slope of each graph represents the rate of temperature change.

17. Sample answer: The slope is less steep (the line should be horizontal) when the water starts to boil. The slope is different because the energy added to the water through heating is making steam rather than increasing the temperature.

18. Sample answer: The slope is less steep (the line should be horizontal) when the water starts to freeze. The slope is different because the removal of energy from the water allows crystal structures (ice) to form rather than decreasing the temperature.

19. During a change of state, the temperature of the substance does not change.

20. First graph: When the temperature of the water first starts to increase, the water molecules are moving faster. When the water starts to boil and the temperature levels off, the water molecules are beginning to move fast enough to overcome the attractive forces between them and become a gas. Second graph: As the temperature of the water decreases, the water molecules' motion is slowing down. When the temperature levels off, the water molecules have slowed down enough to lock into place and become a solid, ice.

DATASHEET C

14. The temperature remains constant.

15. The temperature remains constant.

16. The slope of each graph represents the rate of temperature change.

17. Sample answer: The slope is less steep (line should be horizontal) when the water starts to boil. The slope is different because the energy added to the water through heating is making steam rather than increasing the temperature.

18. Sample answer: The slope is less steep (line should be horizontal) when the water starts to freeze. The slope is different because the removal of energy from the water allows crystal structures (ice) to form rather than decreasing the temperature.

19. Sample answer: During a change of state, the temperature of the substance does not change.

20. Sample answer: First graph: When the temperature of the water first starts to increase, the water molecules are moving faster. When the water starts to boil and the temperature levels off, the water molecules are beginning to move fast enough to overcome the attractive forces between them and become a gas. Second graph: As the temperature decreases, the water molecules' motion is slowing down. When the temperature levels off, the water molecules have slowed down enough to lock into place and become a solid, ice.

Science Skills Activity

DATASHEET

1. Sample calculation:
100 K/1.0 atm = 100 K/atm
200 K/2.0 atm = 100 K/atm
The same ratio was obtained from the two data points, indicating that the relationship is linear.

2. Students' graphs should resemble the following:

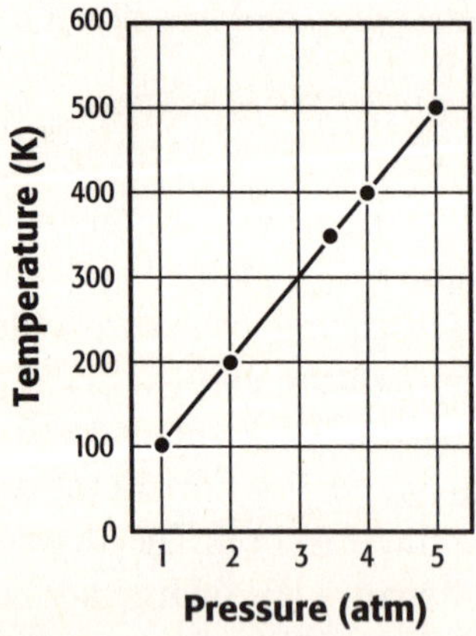

The graph of my data confirms that the relationship is linear because the data points form a straight line when connected.

3. The linear relationship between pressure and temperature indicates that they will vary together at a constant rate.